MICH. DEPT. OF COMMUNITY HEALTH
BUREAU OF LABORATORIES
CHEMISTRY & TOXICOLOGY DIVISION
3500 N. MARTIN LUTHER KING BLVD.
P.O. BOX 30035
LANSING, MI 48909

ANALYTICAL METHOD VALIDATION AND INSTRUMENT PERFORMANCE VERIFICATION

ANALYTICAL METHOD VALIDATION AND INSTRUMENT PERFORMANCE VERIFICATION

Edited by

CHUNG CHOW CHAN
Eli Lilly Canada, Inc.

HERMAN LAM
GlaxoSmithKline Canada, Inc.

Y. C. LEE
Patheon YM, Inc.

XUE-MING ZHANG
Novex Pharma

WILEY-
INTERSCIENCE

A JOHN WILEY & SONS, INC., PUBLICATION

Library of Congress Cataloging-in-Publication Data:

Analytical method validation and instrument performance verification /
Chung Chow Chan ... [et al.].
    p. ; cm.
Includes bibliographical references and index.
  ISBN 0-471-25953-5 (cloth : alk. paper)
  1. Drugs—Analysis—Methodology—Evaluation. 2.
Laboratories—Equipment and supplies—Evaluation. 3.
Laboratories—Instruments—Evaluation.
    [DNLM: 1. Chemistry, Pharmaceutical—instrumentation. 2. Chemistry,
Pharmaceutical—methods. 3. Clinical Laboratory Techniques—standards.
4. Technology, Pharmaceutical—methods. QV 744 A532 2004] I. Chan,
Chung Chow.
  RS189.A568 2004
  610'.28—dc21

                  2003014141

CONTENTS

CONTRIBUTORS

Nicole E. Baryla, Ph.D., Eli Lilly Canada, Inc., 3650 Danforth Avenue, Toronto, Ontario M1N 2E8, Canada

Heiko Brunner, Ph.D., Lilly Forschung GmbH, Essener Strasse 93, D-22419 Hamburg, Germany

Chung Chow Chan, Ph.D., Eli Lilly Canada, Inc., 3650 Danforth Avenue, Toronto, Ontario M1N 2E8, Canada

Fabio Garofolo, Ph.D., Vicuron Pharmaceuticals, Inc., via R. Lepetit 34, I-21040 Gerenzano, Italy

Ludwig Huber, Ph.D., Agilent Technologies, Hewlett-Packard Strasse 8, 76337 Waldbronn, Germany

Chantal Incledon, GlaxoSmithKline Canada, Inc., 7333 Mississauga Road North, Mississauga, Ontario L5N 6L4, Canada

Rick Jairam, GlaxoSmithKline Canada, Inc., 7333 Mississauga Road North, Mississauga, Ontario L5N 6L4, Canada

Eric Jensen, Ph.D., Eli Lilly & Company, Indianapolis, IN

Herman Lam, Ph.D., GlaxoSmithKline Canada, Inc., 7333 Mississauga Road North, Mississauga, Ontario L5N 6L4, Canada

Y.C. Lee, Ph.D., Patheon YM, Inc., 865 York Mills Road, Toronto, Ontario M3B 1Y5, Canada

Robert Metcalfe, Ph.D., GlaxoSmithKline Canada, Inc., 7333 Mississauga Road North, Mississauga, Ontario L5N 6L4, Canada

Yoshiki Nishiyama, Eli Lilly Japan KK, 4-3-3 Takatsukadai, Nishi-ku, Kobe 651-2271, Japan

Neil Pearson, Eli Lilly Canada, Inc., 3650 Danforth Avenue, Toronto, Ontario M1N 2E8, Canada

Anna Rebelo-Cameirao, Eli Lilly Canada, Inc., 3650 Danforth Avenue, Toronto, Ontario M1N 2E8, Canada

Yu-Hong Tse, Ph.D., GlaxoSmithKline Canada, Inc., 7333 Mississauga Road North, Mississauga, Ontario L5N 6L4, Canada

Gilman Wong, GlaxoSmithKline Canada, Inc., 7333 Mississauga Road North, Mississauga, Ontario L5N 6L4, Canada

Xue-Ming Zhang, Ph.D., Novex Pharma, 380 Elgin Mills Road East, Richmond Hill, Ontario L4C 5H2, Canada

PREFACE

For pharmaceutical manufacturers to achieve commercial production of safe and effective medications requires the generation of a vast amount of reliable data during the development of each product. To ensure that reliable data are generated in compliance with current Good Manufacturing Practices (cGMPs), all analytical activities involved in the process need to follow Good Analytical Practices (GAPs). GAPs can be considered as the culmination of a three-pronged approach to data generation and management: method validation, calibrated instrument, and training. The requirement for the generation of reliable data is very clearly represented in the front cover design, where the three strong pillars represent method validation, calibrated instrument, and training, respectively.

This book is designed to cover two of the three pillars of data generation. The chapters are written with a unique practical approach to method validation and instrument performance verification. Each chapter begins with general requirements and is followed by the strategies and steps taken to perform these activities. The chapter ends with the author sharing important practical problems and their solutions with the reader. I encourage you to share your experience with us, too. If you have observations or problem solutions, please do not hesitate to email them to me at chung_chow_chan@cvg.ca. With the support of the Calibration & Validation Group (CVG) in Canada, I have set up a technical solution-sharing page at the Web site *www.cvg.ca*. The third pillar, training, is best left to individual organizations, as it will be individualized according to each organization's strategy and culture.

The method validation section of this book discusses and provides guidance for the validation of common and not-so-common analytical methods that are used to support development and for product release. Chapter 1 gives an overview of the activities from the discovery of new molecules to the launch of new products in the pharmaceutical industry. It also provides an insight into quality systems that need to be built into the fundamental activities of the discovery and development processes. Chapters 2 to 5 provide guidance and share practical information for validation of common analytical methods (e.g., potency, related substances, and dissolution testing). Method validation for pharmaceutical excipients, heavy metals, and bioanalysis are discussed in Chapters 6 to 8.

The instrument performance verification section of the book provides unbiased information on the principles involved in verifying the performance of instruments that are used for the generation of reliable data in compliance with cGMPs. The reader is given different approaches to the successful verification of instrument performance. The choice of which approach to implement is left to the reader based on the needs of the laboratory. Chapters 9 to 15 provide information on common analytical instruments used in the development laboratory (e.g., HPLC, UV–Vis spectrophotometers, and pH meters). Chapter 13 provides a detailed discussion of the LC-MS system, which is fast becoming a standard analytical laboratory instrument. Since a great portion of analytical data from the drug development process comes from stability studies, Chapter 16 is included to provide guidance to ensure proper environmental chamber qualification.

Computers have become a central part of the analytical laboratory. Therefore, we have dedicated the last two chapters to an introduction to this field of computer system and software validation. Chapter 17 guides quality assurance managers, lab managers, information technology personnel, and users of equipment, hardware, and software through the entire qualification and validation process, from writing specifications and vendor qualification to installation and to both initial and ongoing operations. Chapter 18 is an in-depth discussion of the approaches to validation of Excel spreadsheets, one of the most commonly used computer programs for automatic or semiautomatic calculation and visualization of data.

The authors of this book come from a broad cultural and geographical base of pharmaceutical companies, vendors and contract manufacturers and offer a broad perspective to the topics. I want to thank all the authors, co-editors, reviewers, and the management teams of Eli Lilly & Company, GlaxoSmithKline Canada, Inc., Patheon Canada, Inc., Novex Pharma, and Agilent Technologies who have contributed to the preparation of this book. In addition, I want to acknowledge Herman Lam for the design of the front cover, which clearly depicts the cGMP requirements for data generation.

CHUNG CHOW CHAN, PH.D.

1

OVERVIEW OF PHARMACEUTICAL PRODUCT DEVELOPMENT AND ITS ASSOCIATED QUALITY SYSTEM

CHUNG CHOW CHAN, PH.D.
Eli Lilly Canada, Inc.

ERIC JENSEN, PH.D.
Eli Lilly & Company, Indianapolis

1.1 INTRODUCTION

Pharmaceutical product development consists of a series of logical and systematic processes. When successful, the final outcome is a commercially available dosage form. However, this process can become a long and complicated process if any of the steps lose their focus. The industry has undergone many changes over the years to increase focus on efficiency and efficacy of the development process. The overall cycle of pharmaceutical product development is summarized in Figure 1.1. The clinical study of drug development is the most obvious and best known to laypersons and scientists. However, many associated behind-the-scene activities are also actively pursued in a parallel and timely manner to ensure the success of pharmaceutical product development. Clinical and commercial success cannot be achieved without successful completion of these other activities. It is important to note that the clinical phase boxes in Figure 1.1 may not be aligned exactly chronologically with other development activities.

Analytical Method Validation and Instrument Performance Verification, Edited by Chung Chow Chan, Herman Lam, Y. C. Lee, and Xue-Ming Zhang
ISBN 0-471-25953-5 Copyright © 2004 John Wiley & Sons, Inc.

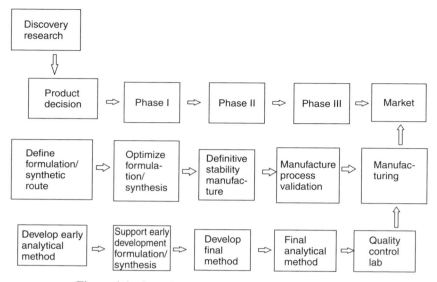

Figure 1.1. Overview of the drug development process.

Historically, the time period for pharmaceutical drug product development is usually on the order of 10 to 15 years. However, with the ever-increasing competition between pharmaceutical companies, it is of utmost important to reduce the time utilized to complete the development process.

1.1.1 Discovery Research

In the discovery research phase of drug development, new compounds are created to meet targeted medical needs, hypotheses for model compounds are proposed, and various scientific leads are utilized to create and design new molecules. Thousands of molecules of similar structure are synthesized to develop a structure–activity relationship (SAR) for the model. To reach this stage, large pharmaceutical companies rely on new technologies, such as combinatorial chemistry and high-throughput screening, which are cornerstones in drug discovery. The new technologies increase the choice of compounds that can be synthesized and screened. Various in vivo and in vitro models are used to determine the value of these new candidate compounds.

The sequencing of the complete human genome was completed in 2000 through the Human Genome Project, which was begun in 1995. Knowledge of the complete human genome will provide the basis for many possible targets for drug discovery through genomics, proteonomics, and bioinformatics.

1.1.2 Preclinical Phase

The most promising drug candidates would be worthless if they could not be developed, marketed, or manufactured. New therapeutic drugs from drug discovery will undergo extensive testing to obtain initial safety and efficacy data in animal models. Upon completion of successful animal safety and efficacy evaluation, submission to appropriate regulatory bodies is made to gain approval to administer the first human dose in the clinical phase I trial.

1.1.3 Clinical Phases

The clinical phase I trial is used to assess the safety and, occasionally, the efficacy of a compound in a few healthy human volunteers. These studies are designed to determine the metabolism and pharmacological action of the drug in humans, the side effects associated with increasing doses, and if possible, to gain very early information on the drug's effectiveness. Safety data from these trials will help determine the dosage required for the next phase of drug development. The total number of subjects in phase I studies is generally in the range 20 to 80.

Clinical phase II trials are conducted to evaluate the effectiveness of a drug for a particular indication or indications in patients with the targeted disease. These studies also help to determine the common short-term side effects and risks associated with the drug. Phase II studies are typically well controlled, closely monitored, and conducted in a relatively small number of patients, usually no more than several hundred.

Active Pharmaceutical Ingredient (API). In this early stage of drug development, only a small quantity of drug substance is needed. As development progresses into later stages, greater quantities of drug substance are needed and will trigger efforts to optimize the synthetic route.

Formulation Development. The formulation of the new drug product will be designed in conjunction with medical and marketing input. Excipients to be used will be tested for chemical and physical compatibility with the drug substance. The preliminary formulation design will be optimized at this stage.

Analytical Development of API and Drug Products. Early methods to support synthetic and formulation developments are often developed in the form of potency assay, impurities/related substance assay, dissolution, Karl Fischer, identity, chiral method, and content uniformity. These analytical methods are developed and validated in a fast and timely manner to support all phase II studies.

Common Studies Performed on the API and Drug Product. At this stage of the development, it is important to gain preliminary information of the stability of the API and drug product. Therefore, open dish (i.e., nonprotected) stability studies are carried out to understand the chemical and physical stability of both the API and the drug product. Preliminary packaging stability studies are conducted to obtain a preliminary assessment of packaging materials that can be used, and photostability and thermal studies are conducted to determine the light and thermal stability of the API and drug product.

Successful efficacy and safety data will guide the decision to proceed to clinical phase III in product development. In this stage, the new drug is administered to a larger population of patients using blinded clinical studies. These studies may demonstrate the potential advantages of the new compound compared with similar compounds already marketed. The data collected from this stage are intended to evaluate the overall benefit–risk relationship of the drug and to provide an adequate basis for labeling. Phase III studies usually include from several hundred to several thousand subjects and often include single- or double-blind studies designed to eliminate possible bias on the part of both physicians and patients. Positive data from this stage will trigger implementation of a global registration and commercialization of the drug product.

Impurities Level in New Drug Product. As the new drug product formulation progresses to this late stage of development, impurity profiles may differ from those of earlier formulations. The rationale for reporting and control of impurities in the new drug product is often decided at this stage as are recommended storage conditions for the product. Degradation products and those arising from excipient interaction and/or container closure systems will be isolated and identified. The impurity profile of the representative commercial process will be compared with the drug product used in development, and an investigation will be triggered if any difference is observed. Identification of degradation products is required for those that are unusually potent and produce toxic effects at low levels.

Primary and developmental stability studies help development scientists understand the degradation pathways. These studies are developed to get information on the stability of the drug product, expected expiry date, and recommended storage conditions. All specified degradation products, unspecified degradation products, and total degradation products are monitored in these studies.

Impurities in API. Treatment of the impurities in the API is similar to that for the new drug product. Impurities in the API include organic impurities (process and drug related), inorganic impurities, and residual solvents. Quality control analytical procedures are developed and validated to ensure appropriate detection and quantitation of the impurities. Specification limits for impurities are set based on data from stability studies and chemical development studies. A rationale for the inclusion or exclusion of impurities is set at this stage. The limits set should not be above the safety level or below the limit of the manufacturing process and analytical capability.

API Development. The synthetic route will be finalized and a formal primary stability study will be undertaken to assess the stability of the API.

Formulation Development. The formulation is finalized based on the experience gained in the manufacture of clinical phase I and II trial materials. Scale-up of the manufacturing process will be completed to qualify the manufacturing capability of the facility. The primary stability study is initiated to assess the stability of the drug product.

1.1.4 Regulatory Submission

Successful completion of clinical phase III trial is a prerequisite for the final phase of drug development. The complete set of clinical, chemical, and analytical data is documented and submitted for approval by regulatory agencies worldwide. Simultaneous activities are initiated to prepare to market the product once regulatory approval is received. As clinical phase III is still being conducted on a limited number of patients, postmarketing studies (phase IV) are often required by regulatory agencies to ensure that clinical data will still be valid. At this point, the company will initiate information and education programs for physicians, specialists, other health care providers, and patients as to the indications of the new drug.

1.2 QUALITY SYSTEM FOR THE ANALYTICAL DEVELOPMENT LABORATORY

As global regulatory requirements have become more similar as the result of deliberate harmonization, analytical methods for global products must be able to meet global regulatory requirements. Ideally, a method developed and validated in the United States should not need to be revalidated or require patchwork validation for use in Japan or Europe. The achievement of this objective is the responsibility of senior management and requires participation and commitment by personnel in many different functions at all levels within the establishment and by its suppliers. To achieve this objective reliably, there must be a comprehensively designed and correctly implemented system of quality standards incorporating GMPs. It should be fully documented and effectively monitored. All parts of the quality systems should be adequately resourced with qualified personnel and suitable premises, equipment, and facilities. It is our intent in the second part of this chapter to give an overview of the extent and application of analytical quality systems to different stages of the drug development process.

1.2.1 Consideration for Quality Systems in Development

An important consideration in the development of quality systems in development is to ask the question: What business does development support? Development does not mean exclusively working to develop formulation or analytical methods; many activities are directly involved in support of clinical material production. Laboratory leadership has the responsibility to consider carefully the customers and functions of an analytical development department. As part of this consideration, several key questions are useful in defining the business and quality standards:

- How does the larger organization view development?
- How close to discovery is development?
- How close to manufacturing is development?

- Where are there major overlaps in activities and support?
- What is the desirable quality culture for this organization?
- Who are the primary customers of development's outputs?

When a new molecule enters the development phase, in most cases only the basic information of the new chemical entity is known (e.g., molecular structure and polymorphic and salt forms). However, we do not know what will happen when it is formulated and stored at ordinary environmental conditions. In other words, there is a high degree of variability around what is "known" about the molecule and its behavior in a variety of systems. The basic task for development is to reduce this high variability by conducting a series of controlled experiments to make this information known and thus predictable. In fact, by the time a molecule reaches the significant milestone of launch into commercial activities, most of the behavior and characteristics of the molecule need to be known, predictable, and in control.

There are multiple paths to achieving the state when a product and a process are "in control." A pictorial representation of this concept is shown in Figure 1.2. Simpler molecules may achieve a state of control (predictable state) early in the development process, while more complex molecules may retain a high state of "variability" until late in the process. The goal for development must be a development path that is documented and performed by qualified scientists, equipment, facilities, instruments, etc. Development paths that can be followed are varied, but the final outcome, when a project is transferred to manufacturing, is a product and a process that are in a well-characterized state of control.

1.2.2   GMPs Applied to Development

The original intent of the Good Manufacturing Practices (GMPs) was to describe standards and activities designed to ensure the strength, identity, safety, purity, and quality of pharmaceutical products introduced into commerce. Application of GMPs to development activities has evolved to the state where application of

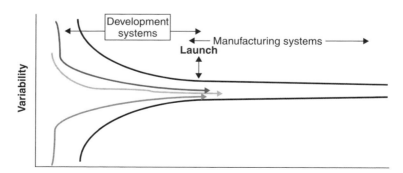

Figure 1.2. Variability during the development process.

the basic GMP principles is a common part of business practice for an increasing number of companies. However, the GMPs are silent on explicit guidance for the development phase in several areas. Thus, companies have been left to make their own determinations as to how to apply GMPs prior to commercial introduction of products. More recently, the European Union (EU) and the International Conference on Harmonization (ICH) have offered a variety of guidances in the development of API. The ICH Q7A GMP Guidance for APIs includes guidance for APIs for use in clinical trials. The EU Guideline Annex 13 provides much more specific guidance to the application of GMPs to investigational medicinal products. By extension, one can gain perspective on application of GMPs to the chemistry, manufacturing, and control (CM&C) development process since it is closely tied to the development, manufacture, and use of investigational medicinal products.

Regulatory bodies recognize that knowledge of the drug product and its analytical methods will evolve through the course of development. This is stated explicitly in ICH Q7A: Changes are expected during development, and every change in production, specifications, or test procedures should be recorded adequately. The fundamental nature of the development process is one of discovery and making predictable the characteristics of the API or product. It is therefore reasonable to expect that changes in testing, processing, packaging, and so on will occur as more is learned about the molecule. A high-quality system that supports development must be designed and implemented in a way that does not impede the natural order of development. It must also ensure that the safety of subjects in clinical testing is not compromised. The penultimate manufacturing processes must be supported with sufficient data and results from the development process so that the final processes will be supported in a state of control.

Processes that are created during development cannot achieve a full state of validation because the processes have not been finalized. Variation is an inherent part of this process, and it allows the development scientists to reach conclusions concerning testing and manufacturing after having examined these processes with rigorous scientific experiments and judgments. The goal for development is to arrive at a state of validation entering manufacturing.

If one looks at the various clinical stages of development, there is a question as to what practices should be in place to support phase I, II, or late phase III studies. An all-or-nothing approach to GMPs is not appropriate. There are certain fundamental concepts that must be applied regardless of the clinical phase of development. Examples of these include: (1) documentation, (2) change, (3) deviations, (4) equipment and utilities, and (5), training.

Any high-quality system must be built with an eye to the regulations and expectations of the regulatory agencies that enforce the system. The U.S. Food and Drug Administration (FDA) has recently implemented a systems approach of inspection for ensuring that current GMPs (cGMPs) are followed in the manufacturing environment. The FDA will now inspect by systems rather than by specific facility or product. The premise in this system is that activities in a pharmaceutical company can be organized into systems that are sets of operations and related activities. Control of all systems helps to ensure that the firm will produce drugs that are safe and that have the identity and strength and meet the quality and purity characteristics that are intended. The goal of this program's activities is to minimize consumers exposure to adulterated drug products. A company is out of control if any of its systems is out of control.

The following six systems are identified to be audited in the FDA systems (GMP subparts are shown in parentheses):

Quality system (B, E, F, G, I, J, K)   Facilities and equipment systems (B, C, D, J)
Materials system (B, E, H, J)   Packaging and labeling systems (B, G, J)
Production system (B, F, J)   Laboratory and control systems (B, I, J, K)

An analysis of the citations of each cGMP system reveals that two subparts are included in all the citations: *Organization and Personnel* (subpart B) and *Records and Reports* (subpart J). This analysis points to a fundamental precept in the systems guidance. Having the right number of appropriately qualified personnel in place along with a strong documentation, records, and reports system are the foundation of success in implementation of cGMPs in a manufacturing operation. It therefore follows that the same principles apply to the development processes that lead to the successful implementation of manufacturing operations.

During the development process, it is important to control variables that affect the quality of the data that are generated and the ability to recreate the work. It is important to recognize that by its nature, the development process does not achieve a complete success rate. That is, many more molecules enter drug development than transit successfully to the market. Thus, it is reasonable to develop guidance and practices as to how much control and effort are put into key activities depending on the phase of development. For example, analytical methods used to determine purity and potency of an experimental API that is very early in development will need a less rigorous method validation exercise than would be required for a quality control laboratory method used in manufacturing. An early phase project may have only a limited number of lots to be tested; the testing may be performed in only one laboratory by a limited number of analysts. The ability of the laboratory to "control" the method and its use is relatively high, particularly if laboratory leadership is clear in its expectations for the performance of the work.

The environment in which this method is used changes significantly when the method is transferred to a quality control laboratory. The method may be replicated in several laboratories, multiple analysts may use it, the method may be one of many methods used in the laboratory, and the technical depth of the analysts may be less deep than those in the development laboratory. Thus, it is incumbent on the development laboratory to recognize when projects move to later phases of development. The developing laboratory must be aware of the needs of the receiving laboratories as well as regulatory expectations for successful validation of a method to be used in support of a commercial product. The validation exercise becomes larger; more detailed, and collects a larger body of data to ensure that the method is robust and appropriate for use.

Similar examples apply to the development of synthetic and biochemical process for generation of API as well as experimental pharmaceutical products. These examples are familiar to scientists who work in the drug development business. Unexpected findings are often part of the development process. A successful quality system that supports this work will aid in the creation of an environment that ensures that this work is performed in an environment where the quality of the data and results are well controlled. Thus, one would expect strong emphasis on documentation systems and standards, employee training and qualification, equipment and instrument qualification, and utilities qualification. Controlling these variables provides a higher degree of assurance in the results and interpretation of results.

The culture around quality within the development business will make or break the success of the quality system of the business. It is important that the development process be described and mapped. The process should be documented and the process understood. The path that the development area will be taking begins with the decision to develop the molecule. Actions are needed to ensure that an appropriate quality system will be implemented and maintained. Financing for the quality system should be given appropriate financial backup to ensure a functional system and not a minimal budget. The culture of the development area in the company should understand the full value of quality. It is wrong to focus solely on speed of development and work with the attitude of fixing quality issues as the process is developed while hoping that any problems that occur will never be found. Quality must include the willingness of development leadership to invest in systems and processes so that development can go rapidly.

It is important to recognize signs in the development laboratory which indicate that the quality system has been implemented successfully. The following list includes some of the observations that can be made easily if the quality system is functioning as intended.

1. Expectations are high for documentation and reports. This observation demonstrates the maturity of the scientists in the laboratory to think and practice good quality principles.
2. Processes for planning and conducting work are robust.
3. Project planning includes quality objectives.
4. The system is able to accommodate all types of molecules.
5. The development process is mapped and followed.
6. Leadership is actively involved.

1.3 CONCLUSIONS

This introductory chapter gave a quick overview of the drug discovery process. Normal activities required from molecule discovery to launch of the product are described. Quality systems for drug development must be built with an eye to the fundamental aspects of the discovery and development processes. There must be recognition that there is evolution on some standards during development. The business must have clarity about its purpose and the processes used to run the business. Quality expectations must be part of the development culture to ensure compliance with cGMP requirements. Quality and business leadership must provide a capable environment in which discovery and development occur.

2

POTENCY METHOD VALIDATION

CHUNG CHOW CHAN, PH.D.
Eli Lilly Canada, Inc.

2.1 INTRODUCTION

Assay as defined by the *Japanese Pharmacopoeia* is a test to determine the composition, the content of the ingredients, and the potency unit of medicine by physical, chemical, or biological procedures. This chapter focuses on validation of the potency assay by high-performance liquid chromatography (HPLC). Analytical method development and validation involve a series of activities that are ongoing during the life cycle of a drug product and drug substance. Figure 2.1 summarizes the life cycle of an analytical method.

Analytical potency method development should be performed to the extent that it is sufficient for its intended purpose. It is important to understand and know the molecular structure of the analyte during the method development process, as this will facilitate the identification of potential degradation impurities. For example, an impurity of $M + 16$ in the mass spectrum of a sample may indicate the probability of a nitrogen oxide formation. Upon successful completion of method development, the potency method will then be validated to show proof that it is suitable for its intended purpose. Finally, the method validated will be transferred to the quality control laboratory in preparation for the launch of the drug substance or drug product.

The method will be used in the manufacturing facility for the release of both drug substance and drug product. However, if there are any changes in the manufacturing process that have the potential to change the degradation pattern

Analytical Method Validation and Instrument Performance Verification, Edited by Chung Chow Chan, Herman Lam, Y. C. Lee, and Xue-Ming Zhang
ISBN 0-471-25953-5 Copyright © 2004 John Wiley & Sons, Inc.

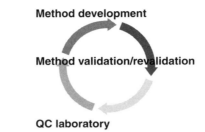

Figure 2.1. Life cycle of an analytical method.

of the drug substance and drug product, this validated method may need to
be revalidated. This process of revalidation is described in more detail later in
the chapter.

Whether it is a drug substance or a drug product, the final product will need
to be analyzed to assess its potency or strength. The potency of a drug substance
is typically reported as a percentage value (e.g., 98.0%), whereas a drug product
is reported in terms of its intended concentration or label claim.

2.2   SCOPE OF CHAPTER

In this chapter we outline the general requirements for the HPLC potency method
validation in pharmaceutical products. The discussion is based on method validation for small-molecule pharmaceutical products of synthetic origin. Even though
most of the requirements are similar for a drug substance, method validation
for a drug substance is not discussed in detail in this chapter. The discussion
focuses on current regulatory requirements in the pharmaceutical industry. Since
the expectations for method validation are different at different stages of the
product development process, the information given in this chapter is most suitable for the final method validation according to International Conference on
Harmonization (ICH) requirements to prepare for regulatory submissions [e.g.,
New Drug Application (NDA)]. Even though the method validation is related
to HPLC analysis, most of the principles are also applicable to other analytical
techniques [e.g., thin-layer chromatography (TLC), ultraviolet analysis (UV)].

ICH Q2A [1] proposed the guidelines shown in Table 2.1 for the validation
of a potency assay for a drug substance or drug product.

In this chapter we discuss the following topics regarding validation practices:

1. Types of quantitation technique
2. System suitability requirements
3. Stability indicating potency assay
4. Strategies and validation characteristics
5. Revalidation

Table 2.1. Guidelines for Drug Potency Assay

Characteristic	Requirement[a]	Characteristic	Requirement[a]
Accuracy	+	Detection limit	−
Precision		Quantitation limit	−
Repeatability	+	Linearity	+
Intermediate precision[b]	+	Range	+
Specificity	+		

[a] +, Signifies that this characteristic is normally evaluated; −, signifies that this characteristic is not normally evaluated.
[b] In cases where reproducibility has been achieved, intermediate precision is not needed.

2.3 VALIDATION PRACTICES

Different approaches may be used to validate the potency method. However, it is important to understand that the objective of validation is to demonstrate that a procedure is suitable for its intended purpose. With this in mind, the scientist will need to determine the extent of validation required. It is advisable to design experimental work such that the appropriate validation characteristics be considered simultaneously to obtain overall knowledge of the capabilities of the analytical procedure.

2.3.1 Types of Quantitation

Quantitation by External Standard. This quantitation technique is the most straightforward. It involves the preparation of one or a series of standard solutions that approximate the concentration of the analyte. Chromatograms of the standard solutions are obtained, and peak heights or areas are plotted as a function of concentration of the analyte. The plot of the data should normally yield a straight line. This is especially true for pharmaceuticals of synthetic origin. Other forms of mathematical treatment can be used but will need to be justified.

There are some potential instrumental sources of error that could occur using this quantitation technique. It is critical to have minimal variability between each independent injection, as the quantitation is based on the comparison of the sample and standard areas. However, the current autosamplers are able to minimize this variability to less than 0.5% relative standard deviation (RSD).

Quantitation by Internal Standard. Quantitation by internal standard provides the highest precision because uncertainties introduced by sample injection are avoided. In this quantitation technique, a known quantity of internal standard is introduced into each sample and standard solutions. As in the external standard quantitation, chromatograms of the standard and sample solutions are integrated to determine peak heights or peak areas. The ratio of the peak height or area of the analyte to an internal standard is determined. The ratios of the standards are plotted as a function of the concentration of the analyte. A plot of the data should normally yield a straight line.

Due to the presence of the internal standard, it is critical to ensure that the analyte peak be separated from the internal standard peak. A minimum of baseline separation (resolution >1.5) of these two peaks is required to give reliable quantitation. In addition, to quantitate the responses of internal standard accurately, the internal standard should be baseline resolved from any significant related substances and should have a peak height or area similar to that of the standard peak.

2.3.2   Standard Plots for Quantitation

In many instances in the pharmaceutical industry, drug products may be manufactured in a variety of strengths (e.g., levothyroxine tablets in strengths of 50, 100, 150, 200, 500, and 750 μg). To develop and validate these potency methods, three strategies may be followed.

Single-Point Calibration.   A method may be developed and validated using only one standard analyte concentration. The standard plot generated is used to assay the complete range of tablet strengths. This strategy should be adopted wherever possible due to the simplicity of standard preparation and minimal work for quantitation of the sample. However, this method will require different extraction and dilution schemes of the various drug product strengths to give the same final concentration that is in the proximity of the one standard analyte concentration.

Multiple-Point Calibration.   Another strategy involves two or more standard concentrations that will bracket the complete range of the drug product strengths. In this strategy it is critical that the standard plots between the two extreme concentration ranges be linear. Therefore, this is a valid calibration method as long as the sample solutions of different strengths are prepared within the concentration range of the calibration curve. Its advantage is that different strengths can utilize different preparation procedures and be more flexible. Its disadvantage is that multiple weighing of standards at different concentrations may give a weighing error.

One Standard Calibration for Each Strength.   The least favored method is to develop and validate using one standard concentration for each product strength. This situation will arise when the analyte does not exhibit linearity within a reasonable concentration range.

2.3.3   System Suitability Requirements for Potency Assay

Prior to injecting a standard solution in creating the standard plot, it is essential to ensure that the system is performing adequately for its intended purpose. This function is fulfilled by the use of a solution of the system suitability. *System suitability*, an integral part of analytical procedures, is based on the concept that equipment, electronics, analytical operations, and samples constitute an integral system that can be evaluated. System suitability test parameters depend on the procedure being validated.

The following notes should be given due consideration when evaluating a system suitability sample.

1. System suitability is a measure of the performance of a given system on a given day within a particular sample analysis set.
2. The main objective of system suitability is to recognize whether or not system operation is adequate given such variability as chromatographic columns, column aging, mobile-phase variations, and variations in instrumentation.
3. System suitability is part of method validation. Experience gained during method development will give insights to help determine the system suitability requirements of the final method. An example is the hydrolysis of acetylsalicylic acid to salicylic acid in acidic media. Separation of this degradation peak from the analyte could be one criterion for the system suitability of an acetylsalicylic acid assay.
4. A system suitability test should be performed in full each time a system is used for an assay. If the system is in continuous use for the same analysis over an extended period, system suitability should be reevaluated at appropriate intervals to ensure that the system is still functioning adequately for its intended use.
5. System suitability should be based on criteria and parameters collected as a group that will be able to define the performance of the system. Some of the common parameters used include precision of repetitive injections (usually five or six), resolution (R), tailing factor (T), number of theoretical plates (N), and capacity factor (k').

2.3.4 Stability Indicating Potency Assay

It is important to realize that the pharmaceutical regulators require that all potency assays be stability indicating. Regulatory guidance in ICH Q2A, Q2B, Q3B, and FDA 21 CFR Section 211 [1–5] all require the development and validation of stability-indicating potency assays. Apart from the regulatory requirements, it is also good scientific practice to understand the interaction of the drug with its physical environment. It is logical and reasonable that the laboratory validate methods that will be able to monitor and resolve degradation products as a result of the stability of the product with the environment. For drug substances, we may need to include synthetic process impurities.

It is common practice to utilize forced degradation studies to accelerate degradation of the drug substance or drug product to get an understanding of its degradation profile. Potential environmental conditions that can be used include 40°C and 75% relative humidity (RH), 50°C and 75% RH, 70°C and 75% RH, or 80°C and 75% RH. Oxidation, reduction, and pH-related degradations are also utilized for degradation studies. Usually, the target is to achieve 10 to 30% degradation. Creating more than 30% degradation will not be useful, due to the potential for secondary degradation. Secondary degradation occurs when the first degradation impurity degrades further. Furthermore, degrading the drug substance or drug product beyond 30% will not be meaningful, since this is unacceptable in the market place.

ICH Q2A suggested validation of the characteristics of accuracy, precision, specificity, linearity, and range for potency and content uniformity assay. A detailed discussion of each of these parameters is presented later in this chapter. Some examples of validation data are presented along with a brief critical discussion of the data.

2.4 STRATEGIES AND VALIDATION PARAMETERS

The most important consideration for strategies of method validation is to design experimental work so that the appropriate validation characteristics are studied simultaneously, thereby minimizing the number of experiments that need to be done. It is therefore important to write some form of protocol to aid the planning process. Executing the experimental work without prior planning will be a disaster for the validation.

2.4.1 Linearity

The ICH defines the *linearity* of an analytical procedure as the ability (within a given range) to obtain test results of variable data (e.g., absorbance and area under the curve) which are directly proportional to the concentration (amount of analyte) in the sample. The data variables that can be used for quantitation of the analyte are the peak areas, peak heights, or the ratio of peak areas (heights) of analyte to internal standard peak. Quantitation of the analyte depends on it obeying Beer's law and is linear over a concentration range. Therefore, the working sample concentration and samples tested for accuracy should be in the linear range.

Linearity is usually demonstrated directly by dilution of a standard stock solution. It is recommended that linearity be performed by serial dilution of a common stock solution. Preparing the different concentrations by using different weights of standard will introduce weighing errors to the study of the linearity of the analyte (in addition to adding more work) but will not help to prove the linearity of the analyte. Linearity is best evaluated by visual inspection of a plot of the signals as a function of analyte concentration. Subsequently, the variable data are generally used to calculate a regression by the least squares method.

As recommended by the ICH, the usual range for the potency assay of a drug substance or a drug product should be ±20% of the target or nominal concentration and ±30% for a content uniformity assay. At least five concentration levels should be used. Under normal circumstances, linearity is achieved when the coefficient of determination (r^2) is ≥ 0.997. The slope, residual sum of squares, and

y-intercept should also be reported as required by the ICH. The slope of the regression line will provide an idea of the sensitivity of the regression and hence the method to be validated. The y-intercept will provide the analyst with an estimate of the variability of the method. For example, the ratio percent of the y-intercept with the variable data at nominal concentration are sometimes used to estimate the method variability. Figures 2.2 and 2.3 illustrate acceptable and nonacceptable linearity data, respectively.

2.4.2 Accuracy

The ICH defines the *accuracy* of an analytical procedure as the closeness of agreement between the values that are accepted either as conventional true values or an accepted reference value and the value found. Accuracy is usually reported as percent recovery by assay, using the proposed analytical procedure, of known amount of analyte added to the sample. The ICH also recommended assessing a minimum of nine determinations over a minimum of

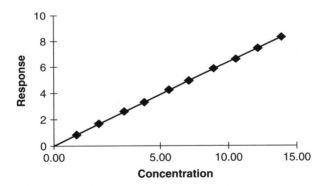

Figure 2.2. Linearity with correlation coefficient greater than 0.997.

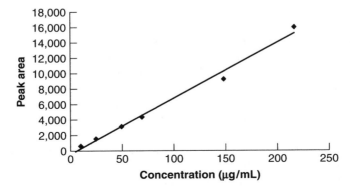

Figure 2.3. Linearity with correlation coefficient less than 0.997.

three concentration levels covering the specified range (e.g., three concentrations/three replicates).

For a drug substance, the common method of determining accuracy is to apply the analytical procedure to the drug substance and to quantitate it against a reference standard of known purity. For the drug product, accuracy is usually determined by application of the analytical procedure to synthetic mixtures of the drug product components or placebo dosage form to which known quantities of drug substance of known purity have been added. The range for the accuracy limit should be within the linear range. Typical accuracy of the recovery of the drug substance in the mixture is expected to be about 98 to 102%. Values of accuracy of the recovery data beyond this range need to be investigated.

2.4.3 Precision

The *precision* of an analytical procedure expresses the closeness of agreement (degree of scatter) between a series of measurements obtained from multiple samples of the same homogeneous sample under prescribed conditions. Precision is usually investigated at three levels: repeatability, intermediate precision, and reproducibility. For simple formulation it is important that precision be determined using authentic homogeneous samples. A justification will be required if a homogeneous sample is not possible and artificially prepared samples or sample solutions are used.

Repeatability (Precision). Repeatability is a measure of the precision under the same operating conditions over a short interval of time. It is sometimes referred to as *intraassay precision.* Two assaying options are allowed by the ICH for investigating repeatability:

1. A minimum of nine determinations covering the specified range for the procedure (e.g., three concentrations/three replicates as in the accuracy experiment), or
2. A minimum of six determinations at 100% of the test concentration.

The standard deviation, relative standard deviation (coefficient of variation), and confidence interval should be reported as required by the ICH.

Tables 2.2 and 2.3 are examples of repeatability data. Table 2.2 shows good repeatability data. However, note that the data show a slight bias below 100% (all data between 97.5 and 99.1%). This may not be an issue, as the true value of the samples and the variation of the assay may be between 97.5 and 99.1%. Table 2.3 shows two sets of data for a formulation at two dose strengths that were performed using sets of six determinations at 100% test concentration. The data indicate a definite bias and high variability for the low-strength dose formulation. It may call into question the appropriateness of the low-dose samples for the validation experiment.

Table 2.2. Repeatability at Different Concentration

Replicate	Concentration (Nominal Concentration 75 µg/mL)					
	14.8	29.6	44.5	74.5	149.1	223.7
1	97.9	98.3	97.6	98.3	98.7	98.6
2	98.0	98.4	97.5	98.2	98.2	98.5
3	98.2	98.5	99.6	99.1	98.1	99.0
4	97.9	98.8	99.1	98.8	97.7	98.8
5	98.4	98.5	97.8	99.0	98.0	99.2
6	99.1	98.6	98.4	98.3	98.1	98.7
Mean	98.3	98.5	98.3	98.6	98.1	98.8
% RSD	0.47	0.18	0.87	0.42	0.33	0.28

Table 2.3. Repeatability at High and Low Concentrations

Replicate	Low Dose	High Dose
1	94.8	100.6
2	91.8	102.1
3	94.1	100.5
4	93.8	99.4
5	95.3	101.4
6	94.7	101.1
Mean	94.1	100.9
% RSD	1.30	0.90

Intermediate Precision. Intermediate precision is defined as the variation within the same laboratory. The extent to which intermediate precision needs to be established depends on the circumstances under which the procedure is intended to be used. Typical parameters that are investigated include day-to-day variation, analyst variation, and equipment variation. Depending on the extent of the study, the use of experimental design is encouraged. Experimental design will minimize the number of experiments that need to be performed. It is important to note that the ICH allows exemption from doing intermediate precision when reproducibility is proven. It is expected that the intermediate precision should show variability that is in the same range or less than repeatability variation. The ICH recommended the reporting of standard deviation, relative standard deviation (coefficient of variation), and confidence interval of the data.

Reproducibility. Reproducibility measures the precision between laboratories as in collaborative studies. This parameter should be considered in the standardization of an analytical procedure (e.g., inclusion of procedures in pharmacopoeias and method transfer between different laboratories). To validate this character-istic, similar studies need to be performed at other laboratories using the same homogeneous sample lot and the same experimental design. In the case of method transfer between two laboratories, different approaches may be taken to achieve the successful transfer of the procedure. However, the most common approach is the *direct method transfer* from the originating laboratory to the receiving labo-ratory. The *originating laboratory* is defined as the laboratory that has developed and validated the analytical method or a laboratory that has previously been cer-tified to perform the procedure and will participate in the method transfer studies. The *receiving laboratory* is defined as the laboratory to which the analytical pro-cedure will be transferred and that will participate in the method transfer studies. In direct method transfer it is recommended that a protocol be initiated with details of the experiments to be performed and acceptance criteria (in terms of the difference between the means of the two laboratories) for passing the method transfer. Table 2.4 gives a set of sample data where the average results obtained between two laboratories were within 0.5%.

2.4.4 Robustness

The *robustness* of an analytical procedure is a measure of its capacity to remain unaffected by small but deliberate variations in the analytical procedure param-eters. The robustness of the analytical procedure provides an indication of its reliability during normal use. The evaluation of robustness should be considered during development of the analytical procedure. If measurements are susceptible to variations in analytical conditions, the analytical conditions should be suitably controlled or a precautionary statement should be included in the procedure. For example, if the resolution of a critical pair of peaks was very sensitive to the percentage of organic composition in the mobile phase, that observation would have been observed during method development and should be stressed in the procedure. Common variations that are investigated for robustness include filter effect, stability of analytical solutions, extraction time during sample prepara-tion, pH variations in the mobile-phase composition, variations in mobile-phase composition, columns, temperature effect, and flow rate.

Table 2.5 shows examples of sample and standard stability performed on an analytical procedure. The two sets of data indicate that the sample and standard

Table 2.4. Results from Method Transfer between Two Laboratories

	Runs	Average %
Originating laboratory	12	100.7
Receiving laboratory	4	100.2

Table 2.5. Stability of Sample and Standard Solutions

Day	% Initial Sample	Standard
1	100.2	99.8
2	100.0	99.8
3	99.9	100.2
4	—	99.5

Table 2.6. Effect of Filter

Replication	Unfiltered	Filter 1	Filter 2
1	101.0	101.1	96.8
2	100.5	101.1	97.1
3	101.0	101.2	96.4
Average	100.8	100.2	96.8

solutions were stable for 3 and 4 days respectively. Table 2.6 gives some data on the effect of a filter on the recovery of the analytical procedure. In a filter study it is common to use the same solution and to compare a filtered solution to an unfiltered solution. For the unfiltered solution, it is common to centrifuge the sample solution and use the supernatant liquid for the analysis. The data set indicated that filter 1 would be recommended for the final analytical procedure.

2.4.5 Specificity

Specificity is the ability to assess unequivocally an analyte in the presence of components that may be expected to be present. In many publications, selectivity and specificity are often used interchangeably. However, there are debates over the use of specificity over selectivity [6]. For the purposes of this chapter, the definition of specificity will be consistent with that of the ICH.

The specificity of a test method is determined by comparing test results from an analysis of samples containing impurities, degradation products, or placebo ingredients with those obtained from an analysis of samples without impurities, degradation products, or placebo ingredients. For the purpose of a stability-indicating assay method, degradation peaks need to be resolved from the drug substance. However, they do not need to be resolved from each other.

Critical separations in chromatography should be investigated at the appropriate level. Specificity can best be demonstrated by the resolution of two chromographic peaks that elute close to each other. In the potency assay, one of the peaks would be the analyte peak. Figure 2.4 illustrates the selectivity of a method to resolve known degradation peaks from the parent peak. Based on the

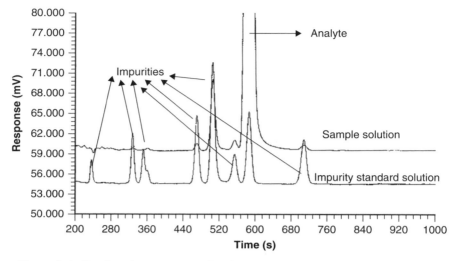

Figure 2.4. Overlay chromatogram of an impurity solution with a sample solution.

experience with the analyte and the chemistry of the analyte, the scientist will be able to identify which of the impurities may be used as the critical pair.

2.5   POTENCY METHOD REVALIDATION

There are various situations during the life cycle of a potency method that require revalidation of the method.

1. During optimization of the formulation or drug substance synthetic process, significant changes may have to be introduced into the process. As a result, to ensure that the analytical method will still be able to analyze the potentially different profile of the drug substance or drug product, revalidation may be necessary.
2. The method was found to be deficient in some areas, such as precision and system suitability. This is especially important as the analytical laboratory gets more experience and more information as to the degradation profile of the sample as it progresses toward submission. If a new impurity is found that makes the method deficient, this method will need to be revalidated.
3. The composition and/or the final manufacturing process of a sample analyzed with the method have been modified after optimization.
4. Changes in equipment or in suppliers of critical supplies at the time of manufacturing. This is important, as critical components of the manufacturing process have the potential to change the degradation profile of the product.

2.6 COMMON PROBLEMS AND SOLUTIONS

In the following pages we summarize some of the common deficiencies of potency method validation. These common problems are grouped together into categories such as HPLC instrumentation, procedural steps, and miscellaneous errors.

2.6.1 HPLC Instrumentation Errors

Qualification of Instruments. The status of the qualification of HPLC and other equipment used for the analytical procedure must always be checked. This is a common error that can lead to reanalysis of the samples if discovered earlier, or repeating the entire experimental procedure if it was discovered after expiry of the sample solutions.

Vacuum Filtering of Mobile Phase. Vacuum filtering of the mobile phase should be avoided in a procedure that is very sensitive to the level of the organic in the mobile phase. Vacuum suction will evaporate the volatile organic portion during filtration (e.g., acetonitrile or methanol), and may lead to variation of the chromatography.

2.6.2 Procedural Errors

Expiry of Mobile Phase. Always check the expiry of mobile phase before use. This is one of the most common errors in an analytical laboratory.

Use of Ion-Pairing Reagents in Mobile Phase. It is usually recommended that if ion-pairing reagents are needed in a mobile phase, its concentration needs to be constant during a gradient run. Changes in ion-pairing concentration during an HPLC run will increase the likelihood of chromatographic variation between runs (e.g., retention time drifts and quantitation precision).

Quantitation of Salts (e.g., Hydrochloride and Sodium Salt). The quantitative result that is reported from the analysis of salts is usually reported with reference to the base of the analyte. The scientist will need to remember to incorporate a multiplier into the calculation to convert the salt data to the base data.

Stability of Standard and Sample Solutions. Appropriate stability of the standard and sample solutions will allow flexibility of the method to be used in a quality control laboratory. For example, 4-day stability of the standard and sample solutions will allow investigation if problems arise during a weekend HPLC run.

Dilution during Sample and Standard Preparation. Minimize the number of dilutions required to give the final dilutions of the sample and standard solutions. Each dilution step will have the potential to introduce error in the procedure.

Range in Validation of Linearity Is Smaller Than Precision and Accuracy. This error will invalidate the precision and accuracy data since the validation did not demonstrate the linearity of the analyte for the quantitation of precision and accuracy data.

2.6.3 Miscellaneous Errors

Validation Protocol. It is highly recommended to validate an analytical procedure using some form of validation protocol. Without a validation protocol, the scientist will have a tendency to vary the experiment during the course of the validation study. Getting into the habit of creating a validation protocol will also ensure that the scientist plans before starting the experiment.

Acceptance Criteria for Validation Parameter. It is highly recommended to set acceptance criteria prior to starting validation experiments. This will provide guidance to the validating scientist on the range of acceptability of the validation results.

Documentation of Observation. It is very important to document all relevant observations during the experimental procedure. Observations are the most important information that can be used if an investigation is needed. Furthermore, observations that are documented provide evidence in the event of patent challenge and other court cases.

Absorbance of Analyte. It is common to devise an experimental procedure that yields an analyte absorbance value of less than 1 absorbance unit. A high absorbance value (depending on the absorptivity of the analyte) is the result of a high concentration of the analyte. Too high a concentration of the analyte may overload the column and lead to nonlinearity.

2.7 SUMMARY OF POTENCY VALIDATION DATA

It is very useful to summarize all method validation data into a tabular format. The tabulated summary will give a quick overview of the validation data. Often, the analyst may be so involved during the actual validation work that some errors escaped detection. Table 2.7 is an example of how data can be recorded.

Table 2.7. Sample Validation Summary

ICH Validation Characteristic	Data Reported	Summary Validation Results
Accuracy	The percent recovery assessed using a minimum of nine determinations over a minimum of three concentration levels covering the range specified.	Based on determinations at three concentration levels, average recovery = 101.4%, RSD = 0.9%.

Table 2.7 (*continued*)

ICH Validation Characteristic	Data Reported	Summary Validation Results
Precision		
Repeatability	The standard deviation, relative standard deviation (RSD), and confidence interval should be reported for each type of precision investigated.	A single experiment ($n = 6$) had a repeatability (RSD) of 1.1%.
Intermediate precision		Based on experimental design ($n = 24$) results from three dosage strengths, the estimated intermediate precision is 0.9%.
Reproducibility		The average potency result obtained from the receiving laboratory is within $\pm$ 0.5% of results from the originating laboratory.
Specificity	Representative chromatograms demonstrate specificity.	The placebo peaks, process impurities, and degradant peaks are resolved from the peak of interest (e.g., Figure 2.4).
Linearity	Data from the regression line (correlation coefficient, y-intercept, slope, residual sum of squares) and a plot.	Correlation coefficient = 0.9999; y-intercept = 0.0328 area unit; slope = 0.5877 [area/(μg/mL)]; residual sum of squares = 178.96 (e.g., data plot in Figure 2.2).
Range	Procedure provides an acceptable degree of linearity, accuracy, and precision when applied to samples containing analyte within or at the extremes of the specified range of procedure.	The range was confirmed as 70 to 130% of the test concentration.

(*continued overleaf*)

Table 2.7 (*continued*)

ICH Validation Characteristic	Data Reported	Summary Validation Results
Robustness	In the case of liquid chromatography, typical variations are: pH in a mobile phase, composition of mobile phase, different columns (different lots and/or suppliers), temperature, and flow rate.	The factors evaluated (analyst, instrument, % ACN, and column age) did not have any significant effect ($p > 0.05$) on the potency results in the ranges studied based on JMP analysis. The method is robust for all the factors studied. The standard and sample solutions were found to be stable for 5 days (at $30°C$).

REFERENCES

1. ICH Harmonized Tripartite Guideline, ICH Q2A, *Text on Validation of Analytical Procedures*, Mar. 1995.
2. CFR Part 211, *Current Good Manufacturing Practice for Finished Pharmaceuticals*.
3. ICH Harmonized Tripartite Guideline, ICH Q2B, *Validation of Analytical Procedures: Methodology*, May 1997.
4. ICH Harmonized Tripartite Guideline, ICH Q3B, *Impurities in New Drug Products*, Oct. 1999.
5. *Drugs Directorate Guidelines*: Health Protection Branch, Health Canada, *Acceptable Methods*, Ottawa, Ontario, Canada, 1994.
6. J. Vessman et al., *Int. Union Pure Appl. Chem. Anal. Chem. Div.* **73**(8), 1381–1386, 2001.

3

METHOD VALIDATION FOR HPLC ANALYSIS OF RELATED SUBSTANCES IN PHARMACEUTICAL DRUG PRODUCTS

Y. C. LEE, PH.D.
Patheon YM, Inc.

3.1 INTRODUCTION

In this chapter we outline the general requirements for analytical method validation for HPLC analysis of related substances in pharmaceutical products. Most
of the discussion is based on method validation for pharmaceutical products of
synthetic origin. Even though most of the requirements are similar for other types
of pharmaceutical drug products (e.g., biopharmaceutical drug products), detailed
discussion of method validation for other types of pharmaceutical drug products
is outside the scope of this chapter. The discussion focuses on current regulatory
requirements in the pharmaceutical industry. Since the expectations for method
validation are different at different stages of the product development process,
the information given in this chapter is most suitable for final method validation according to the ICH requirements to prepare for regulatory submissions
(e.g., NDA). Even though the method validation is related to HPLC analysis,
most of the principles are also applicable to other analytical techniques (e.g.,
TLC, UV).

Analytical Method Validation and Instrument Performance Verification, Edited by Chung Chow
Chan, Herman Lam, Y. C. Lee, and Xue-Ming Zhang
ISBN 0-471-25953-5 Copyright © 2004 John Wiley & Sons, Inc.

3.2 BACKGROUND INFORMATION

3.2.1 Definitions

Definitions for some of the commonly used terms in this chapter are given below.

- *Drug substance* (active pharmaceutical ingredient)*:* a pharmaceutical active ingredient.
- *Related substances:* impurities derived from the drug substance and therefore not including impurities from excipients. Related substances include degradation products, synthetic impurities of drug substance, and manufacturing process impurities from the drug product.
- *Authentic sample:* a purified and characterized sample of a related substance. Unlike reference standards, authentic samples may not be of high purity. However, the purity of an authentic sample has to be determined before use. Authentic samples are used in method development to identify related substances in the analysis. In addition, they are used extensively to prepare the spiked samples in method validation.
- *Spiked sample:* a sample added with a known amount of related substances, prepared from authentic samples during method development or validation.
- *Control sample:* a representative batch of drug substance (or drug product). Typically, control samples are tested in all analyses to ensure consistency in method performance across different runs. Sometimes, they are used as part of the system suitability test to establish the run-to-run precision (e.g., intermediate precision, reproducibility).
- *Response factor:* the response of drug substance or related substances per unit weight. Typically, the response factor of drug substance (or related substance) can be calculated by the following equation:

$$\text{Response factor} = \frac{\text{response(in response units)}}{\text{concentration(in mg/mL)}}$$

- *Relative response factor:* the ratio of the response factor of individual related substance to that of a drug substance to correct for differences in the response of related substances and that of the drug substance. It can be determined using the following equation:

$$\text{Relative response factor} = \frac{\text{response factor of individual related substance}}{\text{response factor of drug substance}}$$

If a linearity curve (Figure 3.1) is constructed for both the related substance and the drug substance by plotting the response versus the concentration, the relative response factor can also be determined by

$$\text{Relative response factor} = \frac{\text{slope}_{\text{related substance}}}{\text{slope}_{\text{drug substance}}}$$

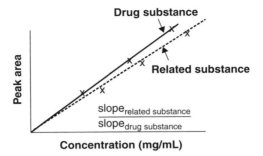

Figure 3.1. Relative response factor.

3.2.2 Different Types of Related Substance Analysis

Area Percent. In this approach, the level of an individual related substance is calculated by the following equation:

$$\%\text{related substance} = \frac{\text{area}_{\text{related substance}}}{\text{total area}} \times 100\%$$

where the $\text{area}_{\text{related substance}}$ is the peak area of the individual related substance and the total area is the peak area (i.e., response) of the drug substance plus the peak areas of all related substances. This is one of the simplest approaches for related substance analysis because there is no need for a reference standard. This is particularly important during the early phase of the project when a highly purified reference standard is not available. It is the preferred approach as long as the method performance meets the criteria described below.

Linearity over a Wide Range of Concentration. Since the areas of the related substances (typically, less than 1%) and drug substance (typically, more than 95%) are summed, it is important that the method is linear from the concentration of related substances (e.g., 1%) to that of the drug substance (e.g., 95%). However, in some cases, the peak shape of the drug substance may not be totally symmetrical at such a high concentration. Therefore, the response may not be linear in such a wide concentration range, and the use of area percentage may not be appropriate. If the response of the analyte is nonlinear at higher concentrations, the related substances would be overestimated. Although this is conservative from a safety perspective, it is inaccurate and therefore unacceptable.

Sample Concentration (Method Sensitivity). To maintain linearity at the concentration range of the drug substance, scientists may try to lower the sample concentration to improve peak shape for the drug substance. However, if the sample concentration is too low, it will affect the method sensitivity, and the ability to detect low levels of related substances may not be adequate.

Response Factor. The response factors of the related substances should be similar to that of the drug substance (i.e., relative response factors close to unity). Otherwise, a response factor correction must be used in the calculation.

High–Low. This approach can be used to overcome the limitation of linear range in the area percent method discussed above. In this approach, samples are prepared at a concentration (i.e., high concentration) similar to that of the area percent method (Figure 3.2). In addition, the high concentration sample solutions are diluted further, to low concentrations (Figure 3.3). Samples from both high- and low-concentration solutions are injected for analysis. In the injections of the high concentration, the responses of all related substances are determined as these small peaks are detectable. The high sample concentration is used to allow all related substances to be detected and quantitated. In the injection of

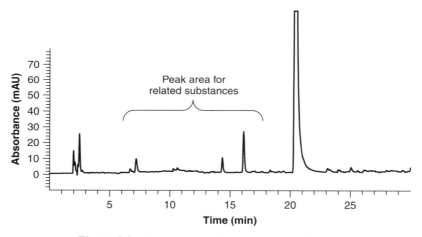

Figure 3.2. chromatogram from high concentration.

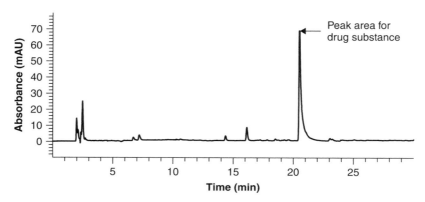

Figure 3.3. chromatogram from low concentration.

low-concentration sample, the response of the drug substance is determined. Low concentration is used to ensure that the response of the drug substance is within the linearity range.

After dilution, response of the drug substance in the low-concentration sample is similar to that of related substance in the high-concentration sample. Therefore, only a small linearity range is required for this method. In addition, since high sample concentration is used for the determination of related substances, high method sensitivity can be achieved. The limitation of the high–low approach is that each sample is injected at least twice (i.e., high and low concentrations) and the total analysis time will be doubled. In addition, an additional step is required to dilute the high concentration to a low concentration, and dilution error can occur during the second dilution.

External Standard. In this approach, related substance levels are determined by calculation using a standard curve. The concentration of related substance is determined by the response (i.e., peak area of individual related substance) and the calibration curve. A reference standard of the drug substance is typically used in the calibration. Therefore, a response factor correction may be required if the response of related substance is very different from that of the drug substance. A single-point standard curve (Figure 3.4) is appropriate when there is no significant *y*-intercept. Otherwise, a multipoint calibration curve (Figure 3.5) has to be used. Different types of calibration are discussed in Section 3.2.3.

The external standard approach offers several advantages over the area percent method, as discussed below.

Reduced Linear Range. Unlike the area percent and high–low methods, which use the response of the drug substance in sample injections for calculation, an external standard method uses a standard curve. Typically, the concentration range of the calibration curve is similar to that of related substances in the sample (e.g., 1 to 5% of the nominal sample concentration). Therefore, this method requires a small linear range.

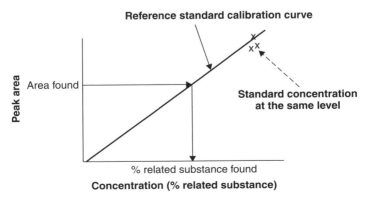

Figure 3.4. Single-point calibration.

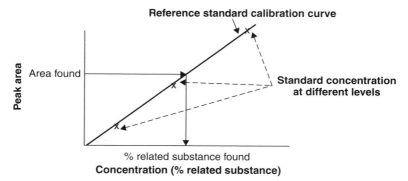

Figure 3.5. Multi-point calibration.

Improved Method Sensitivity. In this approach, only the responses of individual related substances are used in the calculation. Since the area of drug substance peak in the sample injections is not necessary for the calculation, high sample concentrations can be used without worrying about the off-scale response of the drug substance. This approach is particularly useful when the scientists want to improve the method sensitivity by increasing the sample concentration.

Reference Standard. One of the limitations of the external standard method is that a well-characterized reference standard is essential. In addition, each analysis requires accurate weighings of small quantities (e.g., 10 mg) of reference standard. Therefore, weighing error can affect method precision and accuracy.

3.2.3 Suitability of Related Substance Analysis

As discussed in Section 3.2.2, linear range is a critical factor for determining the suitable type of related substance analysis. The following are different situations to illustrate the rationales. Typically, the low end of a linearity curve is about 50% of the ICH reporting limit (e.g., 50% of 0.1% = 0.05%). This is to ensure that the method will be able to calculate results accurately below the ICH reporting limit. The high end of the linearity curve is the nominal concentration (i.e., 100%). This is the target sample concentration for the drug substance.

Case 1. Linearity demonstrated from 50% of the ICH reporting limit to a nominal concentration of drug substance in the sample solution. In addition, no significant *y*-intercept is observed (Figure 3.6). In this case, area percent calculation is suitable because the linearity range covers the responses of related substances and that of the drug substance in the sample solution. Therefore, these responses can be used directly to calculate the area percentage of each related substance.

Case 2. Linearity demonstrated from 50% of the ICH reporting limit to 150% of the shelf life specification of related substance. No significant *y*-intercept is observed (Figure 3.7). In this case, a high–low calculation is more suitable, as

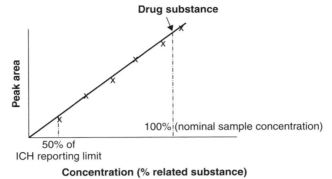

Figure 3.6. Linearity: case 1.

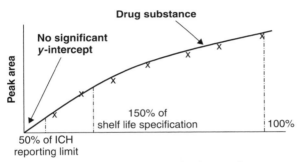

Figure 3.7. Linearity: case 2.

the response is linear only up to the shelf life specification level. Drug substance concentration in sample solution (high concentration) should be diluted to the linear range to obtain the low-concentration solution. Therefore, the response of drug substance in low concentration will be within the linearity range and suitable for calculation. Alternatively, a single-point external standard calibration of concentration within the linearity range can also be used.

Case 3. Linearity demonstrated from 50% of the ICH reporting limit to 150% of the shelf life specification of a related substance, and a significant y-intercept is observed (Figure 3.8). Due to the significant y-intercept, a single-point calibration (e.g., high–low or one-point external standard calibration) is not suitable. In this case, multiple-point external standard calibration is the most appropriate. See Section 3.3.3 for more discussion of the significant y-intercept.

3.2.4 Preparation before Method Validation

Critical Related Substances. Critical related substances are those that may exist at significant levels in the drug product. Authentic samples of these critical related

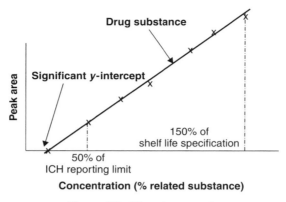

Figure 3.8. Linearity: case 3.

substances should be available for method validation. According to the ICH guidelines, all related substances at a level exceeding the identification threshold have to be identified. These related substances should be considered critical and included in the method validation.

To determine the critical related substances, one can review the related substance profile when the drug substance (or drug product) is subject to stress testing. The most significant related substances in stress testing should be considered critical. In addition, significant related substances (i.e., greater than ICH identification threshold) observed in stability studies during product development should also be included in the method validation. The related substance method has to be validated with respect to each critical related substance; therefore, the workload associated with method validation will increase drastically if the number of critical related substances is large.

Lower and Upper Concentration Range for Method Validation. The concentration range of related substances is typically related to the targeted quantitation limit (QL) at the low end and the proposed shelf life specification at the high end. Therefore, it is important to have a good estimate of these limits; otherwise, inappropriate concentrations may be used in method validation. Even though ICH proposes a method validation range from the ICH reporting limit to 120% of specification, one would want to extend the range to 50% of the ICH reporting limit to 150% of specification to ensure that the method is suitable for most intended uses. The ICH reporting limit is given in Table 3.1. In general, the quantitation limit should be lower than the corresponding ICH reporting limit. This is to ensure that the method is accurate and precise enough to report results at the level of the ICH reporting limit.

Method Procedure. Since the method procedure is undergoing constant modifications during method development, it is very important to define the procedure before method validation. This will ensure that the same method procedure will be used in all method validation experiments.

Table 3.1. Various ICH Thresholds Regarding Degradation Products in New Drug Products as Stated in the Current ICH Guidelines Q3B(R)

Maximum Daily Dose[a]	Threshold[b]
Thresholds for Reporting	
≤1 g	0.1%
>1 g	0.05%
Thresholds for Identification	
<1 mg	1.0% or 5 μg TDI[c] whichever is lower
1–10 mg	0.5% or 20 μg TDI, whichever is lower
>10 mg–2 g	0.2% or 2 mg TDI, whichever is lower
>2g	0.1%
Thresholds for Qualification	
<10 mg	1.0% or 50 μg TDI, whichever is lower
10–100 mg	0.5% or 200 μg TDI, whichever is lower
>100 mg–2 g	0.2% or 2 mg TDI, whichever is lower
>2 g	0.1%

[a] The amount of drug substance administered per day.
[b] Threshold is based on percent of the drug substance.
[c] Total daily intake.

Critical Experimental Parameters for Robustness. Critical experimental parameters should be identified during method development, and they will be investigated in the robustness experiments.

System Suitability Tests. The appropriate system suitability tests should be defined before method validation (e.g., precision, resolution of critical related substances, tailing, detector sensitivity). These system suitability tests should be performed in each method validation experiments. System suitability results from the method validation experiment can be used to determine the appropriate system suitability acceptance criteria.

3.3 METHOD VALIDATION EXPERIMENTS

In this section we outline the requirements for method validation according to current ICH guidelines.

3.3.1   Specificity

ICH definition: Specificity is the ability to assess unequivocally the analyte in the presence of components that may be expected to be present.

Most related substance methods will be used in a stability study, and therefore they have to be stability indicating. *Stability indicating* means that the method has sufficient specificity to resolve all related substances and the drug substance from each other. Typically, for the related substance method for a drug product, degradation products are the most critical related substances. Therefore, as a minimum requirement, the method should have sufficient specificity to resolve the degradation products and the drug substance. In addition, all degradation products should be resolved from potential interference with the excipients.

Samples for Specificity

- Blank solution to show no interference with any HPLC system artifact peak.
- Placebo to demonstrate the lack of interference from excipients.
- Drug substance to show that all significant related substances are resolved from the drug substance.
- Authentic samples of critical related substances to show that all known related substances are resolved from each other.
- Typically, a stressed sample of about 10 to 20% degradation is used to demonstrate the resolution among degradation products. A 10 to 20% degraded sample is used because it has a sufficiently high concentration level of critical related substance. Therefore, these related substances can be detected easily. In addition, 10 to 20% degradation is not too excessive, and the related substance profile should be close to that of a typical stability sample.
- Stressed placebo to show that the degradation products from the excipients will not interfere with the degradation products of the drug substance.

Different Approaches

1. *When authentic samples of related substance are available.* Analyze stressed drug product, placebo, drug substance, stressed placebo, and solutions spiked with authentic samples of related substances. The HPLC chromatograms are used to show the resolution among related substances, drug substance, and other potential interferences. In addition, check the peak homogeneity of the significant degradation products and drug substance by a photodiode array detector (PDA) or mass spectrometer. This verifies that no significant related substance coelute with each other.

2. *When authentic samples of impurities are not available.* A stressed drug product can be analyzed to show separation of the most significant related substances. In addition, the peak homogeneity of the stressed sample should be investigated by PDA or mass spectrometry. Alternatively, one may use an orthogonal procedure to verify the method specificity. The orthogonal

method can be a different technique (e.g., capillary electrophoresis, thin-layer chromatography) or different type of HPLC analysis (e.g., reversed phase versus normal phase). For example, compare the related substance profile in the original HPLC method and that of the orthogonal method. To demonstrate method specificity, the significant related substances should be consistent in these methods.

3.3.2   Quantitation Limit (and/or Detection Limit)

ICH definition: The *quantitation limit* of an individual analytical procedure is the lowest amount of analyte in a sample that can be determined quantitatively with suitable precision and accuracy. The detection limit of an individual analytical procedure is the lowest amount of analyte in a sample that can be detected but not necessarily quantitated as an exact value.

Two types of approaches can be used to determine the quantitation limit or detection limit, as described below.

Signal-to-Noise Approach. *Quantitation limit* (QL; Figure 3.9) is defined as the concentration of related substance in the sample that will give a signal-to-noise (S/N) ratio of 10 : 1. *Detection limit* (DL) corresponds to the concentration that will give a signal-to-noise ratio of 3 : 1. The quantitation limit of a method is affected by both the detector sensitivity and the accuracy of sample preparation at such a low concentration. In practice, the quantitation limit should be lower than the corresponding ICH reporting limit (Table 3.1).

To investigate the effect of both factors (i.e., sample preparation and detector sensitivity), solutions of different concentrations near the ICH reporting limits are prepared by spiking known amounts of related substances into excipients. Each solution is prepared according to the procedure and analyzed repeatedly to determine the S/N ratio. The average S/N ratio from all analyses at each concentration level is used to calculate the QL or DL. The following equation can be used to estimate the QL at each concentration level. Since different concentration levels give different QLs, typically the worst-case QL will be reported as the QL of the method.

$$\text{QL at each concentration} = 10 \times \frac{\text{concentration(in\%related substance)}}{\text{S/N(average at each concentration)}}$$

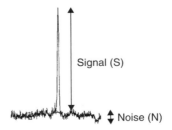

Figure 3.9. Quantitation limit.

Alternatively, the spike solution can be diluted serially to lower concentrations. The S/N ratio at each concentration level is determined. The concentration level (in percent related substance) that gives an S/N value of about 10 will be reported as the QL.

Standard Deviation Approach. The following equations can be used to determine quantitation limit and detection limit by standard deviation of the response at low concentrations:

$$QL = 10 \times \frac{SD}{S}$$

$$DL = 3.3 \times \frac{SD}{S}$$

where SD is the standard deviation of the response near QL and S is the slope of the linearity curve near QL.

There are two ways to determine SD:

1. Using experiments similar to those given for the signal-to-noise approach, determine the standard deviation of the responses by repeat analysis of a solution near the targeted QL.
2. Construct a calibration curve near the targeted QL:
   a. Determine the residual standard deviation of the regression line of calibration, or
   b. Determine the standard deviation of the y-intercept.

Other Considerations for QL. To account for instrument-to-instrument variation, one may need to verify the QL in multiple runs using different instruments. The desired QL should be less than the ICH reporting limit (e.g., 50% of ICH reporting limit). QL should be appropriate; too high indicates that the method is not sensitive enough to report results at the ICH reporting limit. Too low indicates that insignificant degradation products, even though much lower than the ICH reporting limit, may need to be reported.

To ensure that the HPLC system in each analysis is sufficiently sensitive to report results at the ICH reporting limit, one may use a detector sensitivity solution as part of the system suitability test. Since the ICH reporting limit corresponds to QL (i.e., S/N = 10), one-third of the ICH reporting limit should correspond to DL (i.e., S/N = 3). Therefore, as part of the system suitability test, a detector sensitivity solution of a concentration of about one-third of the ICH reporting limit level would be injected. The response of the detector sensitivity solution should meet the detection limit and should be visually distinguishable from baseline.

Alternatively, one may evaluate the S/N ratio of the standard solution during method development or validation. Part of the routine system suitability test is to determine the S/N of the standard solution before each analysis. Therefore, the S/N of each analysis needs to be greater than the established limit.

3.3.3   Linearity

ICH definition: The *linearity* of an analytical procedure is its ability (within a given range) to obtain test results that are directly proportional to the concentration (amount) of analyte in the sample.

General Requirements

Range. Ideally, linearity should be established from 50% of the ICH reporting limit to the nominal concentration of drug substance in the sample solution (for area percent method). If the linearity does not support such a wide range of concentration, determine the linearity from 50% of the ICH reporting level to 150% of the proposed shelf life specifications of the related substance (for the high–low and external standard methods) as a minimum. This will ensure a linear response for related substances at all concentration levels to be detected during stability.

Experimental Requirements. Solutions of known concentrations are used to determine the linearity. A plot of peak area versus concentration (in percent related substance) is used to demonstrate the linearity. Authentic samples of related substances with known purity are used to prepare these solutions. In most cases, for the linearity of a drug product, spiking the related substance authentic sample into excipients is not necessary, as the matrix effect should be investigated in method accuracy.

Acceptance Criteria. Visual inspection is the most sensitive method for detecting nonlinearity. Therefore, the plot has to be linear by visual inspection. In addition, according to ICH guidelines, the following results should be reported: slope, correlation coefficient, *y*-intercept, and residual sum of squares.

y-Intercept. There are several approaches to evaluating the significance of the *y*-intercept.

- *Intercept/slope ratio.* The intercept/slope ratio is used to convert the *y*-intercept from the response unit (peak area) to the unit of percent related substance. The intercept/slope ratio should be compared to the proposed specifications to determine its significance. For example, if the shelf life specification is 2.0%, an intercept/slope ratio of 0.2% may be considered significant, as 0.2% represents 10% relative to the specification.
- *Statistical approach.* The linearity results can be subjected to statistical analysis (e.g., use of statistical analysis in an Excel spreadsheet). The *p*-value of the *y*-intercept can be used to determine if the intercept is statistically significant. In general, when the *p*-value is less than 0.05, the

y-intercept is considered statistically significant. The p-value, which compares the y-intercept with the variation of responses, indicates the probability that the y-intercept to be *not* equal to zero. For example, when the p-value is less than 0.05, this indicates that it is 95% confident that the y-intercept is *not* equal to zero. In other words, it is 95% certain that the y-intercept is significant.

Typically, a positive y-intercept indicates the existence of interference with the response or the saturation of responses at high concentrations. A negative y-intercept indicates the possibility of method sensitivity problem (i.e., a low response cannot be detected) or analytes get retained in the glassware or HPLC system (i.e., a compatibility issue between sample solvent and mobile phase).

Different Approaches for Linearity Determination. The first approach is to weigh different amounts of authentic sample directly to prepare linearity solutions of different concentrations. Since solutions of different concentration are prepared separately from different weights, if the related substances reach their solubility limit, they will not be completely dissolved and will be shown as a nonlinear response in the plot. However, this is not suitable to prepare solutions of very low concentration, as the weighing error will be relatively high at such a low concentration. In general, this approach will be affected significantly by weighing error in the preparation.

Another approach is to prepare a stock solution of high concentration, then perform serial dilution from the stock solution to obtain solutions of lower concentrations for linearity determination. This is a more popular approach, as serial dilution can be used to prepare solutions of very low concentrations. Since the low concentrations are prepared by serial dilution, this approach does not need to weigh a very small quantity of related substance. In addition, since all solutions are diluted from the same stock solution, weighing error in preparing the stock solution will not affect the linearity determination.

Relative Response Factor. The relative response factor (RRF) can be used to correct for differences in relative response between the related substances and the drug substance. In the area percent and high–low method, the related substances are calculated against the response of the drug substance. In the external standard calculation, the standard curve of drug substance is generally used in the calculation. Since the related substances are calibrated by the response of the drug substance, it is necessary to determine the relative response of the related substance to that of the drug substance. After the linearity of the related substances and the drug substance are determined, one can calculate the relative response factor by comparing the slope of the related substance to that of the drug substance. If the relative response factor is significantly different from unity, a correction factor may need to be used in the calculation. Otherwise, the reported results will be grossly over- or underestimated (Figure 3.10).

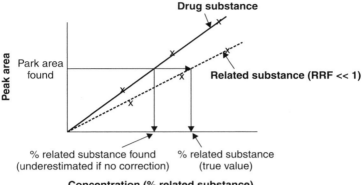

Figure 3.10. Impact of relative response factor.

3.3.4 Accuracy

Definition: The accuracy of an analytical procedure expresses the closeness of agreement between the value that is accepted either as a conventional true value or as an accepted reference value and the value found.

General Considerations

Range. Accuracy for the area percent method should be established from 50% of the ICH reporting limit to the nominal concentration of drug substance in the sample solution. For the high–low and external standard methods, determine accuracy from 50% of the ICH reporting level to 150% of the proposed shelf life specification of the related substances. In addition, for the area percent and high–low methods, it is necessary to determine the accuracy of the related substances and the drug substance. For the external standard method, only the accuracy of related substances is required. Since the response of the drug substance in the sample solution is not used in the external standard calculation, it is not necessary to determine accuracy for the drug substance.

Experimental Considerations. Typically, known amounts of related substances and the drug substance in placebo are spiked to prepare an *accuracy sample* of known concentration of related substance. According to the ICH, accuracy should be determined using a minimum of nine determinations over a minimum of three concentration levels covering the range specified. Similar to the experiments used in linearity (see Section 3.3.3) the related substances and the drug substance can be weighed directly into the placebo or serial dilution from a stock solution can be used. When no authentic sample is available for preparing the spiked solutions, one may determine method accuracy by comparing the results of the original method with that of the orthogonal method (e.g., capillary electrophoresis, thin-layer chromatography, normal-phase HPLC).

Intrinsic Accuracy. Intrinsic accuracy indicates the bias caused by sample matrix and sample preparation. In this approach, a stock solution is prepared by using known quantities of related substance and drug substance. The stock solution is further diluted to obtained solutions of lower concentrations. These solutions are used to generate linearity results. In addition, these linearity solutions of different concentrations are spiked into placebo. The spiked solutions are prepared according to the procedure for sample analysis. The resulting solutions, prepared from the spiked solution, are then analyzed. If the same stock solution is used for both linearity and accuracy and all of these solutions are analyzed on the same HPLC run, the response of linearity (without spike into matrix) and accuracy (with spike into matrix) can be compared directly. Any differences in response indicate the bias caused by matrix interference or sample preparation. To determine the intrinsic accuracy at each concentration level, one can compare the peak area of accuracy (with matrix) with that of linearity (without matrix) at the same concentration (Figure 3.11). This is the simplest approach, and one would expect close to 100% accuracy at all concentration levels.

Overall Accuracy. In addition to the matrix effect and sample preparation error, method accuracy can also be affected by calculation error: for example, difference in relative response, significant y-intercept, and nonlinearity. Therefore, a more vigorous approach is to determine the overall accuracy, which incorporates the effect from all aspects of the method:

- Matrix effect
- Sample preparation
- Calculation error

In this approach, similar to the determination of intrinsic accuracy, the accuracy solutions are prepared by spiking a known quantity of authentic samples and drug

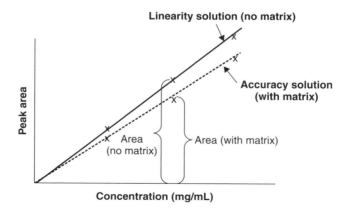

Figure 3.11. Intrinsic accuracy.

substance into excipients. The accuracy solutions are then analyzed using HPLC. The percent related substance results are calculated by the proposed method calculation procedure (e.g., high–low, area percent, external standard). Any correction factors are applied according to the procedure proposed. The overall accuracy is determined by the following equation (Figure 3.12):

$$\text{overall accuracy} = \frac{\%\text{relative substance(calculated)}}{\%\text{relative substance (theory)}}$$

This is a more stringent approach, as this indicates the bias caused by matrix interference, sample preparation, and calculation. For example, related substance (found) $= 1.20\%$ and related substance (theory) $= 1.40\%$ (calculated from the weight of authentic sample used in the spiked solution); therefore,

$$\text{overall accuracy} = \frac{1.20}{1.40} \times 100\% = 86\%$$

3.3.5 Precision

Repeatability. *ICH definition: Repeatability* expresses the precision under the same operating conditions over a short interval of time. Repeatability is also termed *intraassay precision.*

Repeatability of a method can be determined by multiple replicate preparations of the same sample. This can be done either by multiple sample preparations ($n = 6$) in the same experiment or by preparing three replicates at three different concentrations. In general, one should evaluate results of individual related substances, total related substances, and the consistency of related substance profiles in all experiments. The percent RSD and confidence level of these results are reported to illustrate the method repeatability.

Typically, an aged sample should be used to ensure that there are significant levels of related substance in the sample. Alternatively, if different samples are available with different levels of related substance (e.g., fresh sample and sample

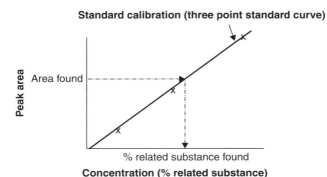

Figure 3.12. Overall accuracy (external standard method).

at expiry), one can determine the repeatability by performing three replicate preparations for each sample. ICH guidelines require a minimum of three samples with three different levels of related substance.

Instead of using spike samples (as in accuracy determination), drug product lots that are representative of the commercial products should be used for precision (repeatability, intermediate precision). This is to ensure that the commercial drug product is used in at least one part of the method validation and that the repeatability results are representative of those that can be expected in the future.

Intermediate Precision. *ICH definition: Intermediate precision* expresses, within laboratories variations, different days, different analysts, different equipment, and so on.

Intermediate precision is to determine method precision in different experiments using different analysts and/or instrument setup. Similar to that of repeatability, one should evaluate the results of individual related substances, total related substances, and the consistency of related substance profiles in all experiments. The percent RSD and confidence level of these results are reported to illustrate the intermediate precision.

Reproducibility. *ICH definition: Reproducibility* expresses the precision between laboratories (collaborative studies are generally used, for standardization of methodology).

This is an optional validation parameter that requires demonstration of laboratory-to-laboratory variation only if multiple laboratories use the same procedure. The reproducibility data can be obtained during method transfer between laboratories.

3.3.6   Range

ICH definition: The *range* of an analytical procedure is the interval between the upper and lower concentrations (amounts) of analytes in the sample (including these concentrations) for which it has been demonstrated that the analytical procedure has a suitable level of precision, accuracy, and linearity (Figure 3.13).

Typically, linearity and accuracy determination covers a wide concentration range (e.g., 50% of the ICH reporting limit to 150% of specification). However, the concentration range for precision will be limited by the availability of sample of different related substance levels. Therefore, to ensure an appropriate method validation range with respect to precision, it is critical to use samples of low and high levels of related substance in precision experiments (e.g., fresh and stressed samples).

3.3.7   Robustness

ICH definition: The *robustness* of an analytical procedure is a measure of its capacity to remain unaffected by small but deliberate variations in method parameters and provides an indication of its reliability during normal use.

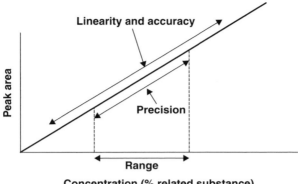

Figure 3.13. Range.

General Considerations. This is to verify that the method performance is not affected by typical changes in normal experiments. Therefore, the variation in method conditions for robustness should be small and reflect typical day-to-day variation. Experimental design (e.g., Plackett–Burman screening, factorial design) is very useful to investigate multiple parameters simultaneously. Critical parameters are identified during the method development process. Only these critical method parameters should be investigated for robustness. Common critical method parameters can be divided into two categories:

1. *HPLC conditions*
    a. HPLC column (lot, age, brand)
    b. Mobile-phase composition (pH ± 0.05 unit, percent organic ± 2%)
    c. HPLC instrument (dwell volume, detection wavelength ± 2 nm, column temperature ± 5°C, flow rate)
2. *Sample preparation*
    a. Sample solvent (pH ± 0.05 unit, percent organic ± 2%)
    b. Sample preparation procedure (shaking time, different membrane filters)
    c. HPLC solution stability

Other Considerations. Typically, the variations in robustness results are compared to the intermediate precision results to demonstrate that robustness is not affected significantly within normal day-to-day variation. When the related substance results are affected by some critical experimental parameters, a precautionary statement needs to be included in the procedure to ensure that this parameter is tightly controlled between experiments. For example, if percent organic of mobile phase affects the results significantly, the procedure should indicate the acceptable range for percent organic (e.g., 50% organic ± 2%)

Built-in Robustness in Method Procedure. The following are some suggestions to improve method robustness:

- *Weighing error.* Weighing error is usually the main source of error. Analytical procedure should ask for a weighing sample or standard of more than 10 mg to minimize weighing error. In addition, use two to three independent weighings in the standard curve and verify the nominal responses of these standard preparations to ensure that there is no significant weighing error. Alternatively, use area percent calculation to eliminate the need for weighing a small quantity of reference standard.
- *Dilution error.* Pipette a volume of more than 5 mL, and avoid using volumetric flasks of less than 25 mL.
- *Sonication.* The efficiency of sonication is highly variable and depends on various factors (e.g., condition of the sonication bath, level of water, and position of flask in the sonication bath). Mechanical shaking is recommended, instead, and is much more reproducible.
- *Mobile phase as sample solvent.* If possible, always use mobile phase as the sample solvent. This ensures the composition (e.g., percent organic, pH) of sample solution matches that of mobile phase and reduces the chance of any problem due to incompatibility of sample solvent and mobile phase. Alternatively, always use sample solvent weaker than that of the mobile phase to ensure that the chromatography is not deteriorated. For example, in reversed-phase HPLC, use less organic solvent in the sample solvent than in the mobile phase.
- *Buffer.* Ensure that the buffer (pK_a) is appropriate for the pH of the solution. In general, to provide appropriate buffering capacity, the pH of the solution should be within ± 1 pH unit of the pK_a value of the buffer.
- *Isocratic method.* Whenever possible, use isocratic HPLC condition, as this is affected less by the variation in flow rate, temperature, and dwell volume. If gradient HPLC conditions have to be used, a simple linear gradient is preferred over multistep gradients. Complicated gradient conditions are more susceptible to differences between HPLC instruments (e.g., flow rate, dwell volume).

3.4 COMMON PROBLEMS AND SOLUTIONS

1. *Presentation of method validation data.* Table 3.2 provides a quick overview of the validation data.

2. *System suitability.* During the robustness testing of method validation, critical method parameters such as mobile phase composition and column temperature are varied to mimic the day-to-day variability. Therefore, the system suitability results from these robustness experiments should reflect the expected range for the system suitability results. As a result, system suitability results in these method validation experiments are very useful in determining the system suitability

Table 3.2. Summary of Validation Results

ICH Validation Characteristic	Analysis	Validation Results
Specificity	Representative chromatograms to demonstrate specificity.	All drug substances and major related substances (A and B) are resolved from each other. There is no significant interference from excipients.
Linearity	Data from regression analysis (correlation coefficient, y-intercept, slope, residual sum of squares) and plot. Linearity is evaluated for A from 0.05 to 1.0% and for B from 0.05 to 2.0% of the nominal sample concentration.	Data from regression analysis:

Compound	A	B
Correlation coefficient	1.000	1.000
y-intercept (% rel. sub.)	0.022	−0.013
Slope (area units/% rel. sub.)	87.2	275.3
Residual sum of squares	177.8	1290.7

ICH Validation Characteristic	Analysis	Validation Results
Range	The procedure provides an acceptable degree of linearity, accuracy, and precision when applied to samples containing analytes within or at the extremes of the specified range of procedure.	The validated range for A is 0.1 to 0.4%, and B is 0.05 to 1.0%.
Accuracy	Assessed using nine determinations over three concentration levels covering the range from 0.1 to 0.4% for A and 0.05 to 1.0% for B.	

Compound	% Accuracy
A	90–100%
B	92–105%

Precision — The average and standard deviation for the individual and total related substances (TRSs) for each drug substance are reported for each type of precision investigated. The overall method precision was evaluated using a combined variance component analysis.

Repeatability

	Average ($n = 16$) (%)	Std. Dev. (%)
A	0.11	0.001
B	0.08	0.002
TRSs	0.18	0.013

(continued overleaf)

Table 3.2. (*continued*)

ICH Validation Characteristic	Analysis	Validation Results		
Intermediate precision			Average ($n = 16$) (%)	Std. Dev. (%)
		A	0.11	0.005
		B	0.08	0.020
		TRS	0.15	0.030
Detection limit (DL) and quantitation limit (QL)	Based on the peak area from the diluted solution of related substances in the sample matrix, the detection and quantitation limit are calculated from the following equations: DL (as % of nominal sample conc.) $= \dfrac{(3 \times \text{std. dev.})}{\text{slope}}$ QL (as % of nominal sample conc.) $= \dfrac{(10 \times \text{std. dev.})}{\text{slope}}$	DL and QL were determined to be 0.004 and 0.02%, respectively. The QL (0.02%) is lower than the corresponding ICH reporting threshold (0.1%).		
Robustness	Typical variations in mobile phase: pH, organic composition, SDS[a] and EDTA[a] concentrations. Typical variations in sample preparation: pH, organic composition, SDS concentration, sample size, sample treatment, and EDTA concentration.	Results from robustness experiments were analyzed by statistical analyses. Variations of all experimental parameters have no significant effect on the procedure based on analysis of the main effects of the factors evaluated for mobile phase and sample preparation.		
	Solution stability of standards and samples were assessed at 5°C and at room temperature.	The sample solutions are stable for 4.5 hours at room temperature and 50 hours (approximately 2 days) at 5°C. The standard solutions are stable for 24 hours at room temperature and 7 days at 5°C.		

[a] SDS, sodium dodecyl sulfate; EDTA, ethylenediaminetetracetic acid.

acceptance criteria. This is a very effective approach since the required system suitability results can be generated during method validation and no other special study is required. However, these results reflect the expected performance of the system, but not necessarily the minimum performance standard for acceptable results. For example, the minimum resolution of the critical pair from method validation may be 3.5; however, a resolution of 2.0 may still be acceptable as long as they are baseline resolved and all other chromatographic parameters remain acceptable.

REFERENCES

1. FDA Guidance for Industry; *Analytical Procedures and Methods Validation* (draft), 2000.
2. ICH Harmonized Tripartite Guideline, ICH Q2A, *Text on Validation of Analytical Procedures*, Mar. 1995.
3. ICH Harmonized Tripartite Guideline ICH Q2B, *Validation of Analytical Procedures: Methodology*, May 1997.
4. ICH Harmonized Tripartite Guideline ICH Q3B(R), *Impurities in New Drug Products*, Oct. 1999.
5. *United States Pharmacopoeia*, USP 25, Chapter <1225>, Validation of Compendial Methods.
6. Snyder, J. Kirkland, and J. Glajch, *Practical HPLC Method Development*, 2nd ed., Wiley, New York 1997.

4

DISSOLUTION METHOD VALIDATION

Chung Chow Chan, Ph.D., Neil Pearson,
and Anna Rebelo-Cameirao
Eli Lilly Canada, Inc.

Y. C. Lee, Ph.D.
Patheon YM, Inc.

4.1 INTRODUCTION

Dissolution testing is one of the most common analytical techniques performed in a pharmaceutical analytical laboratory. It is performed primarily on oral dosage forms to determine the in vitro release of a drug from its finished dosage. Dissolution testing complements other analytical tests that are used to characterize the performance of the final dosage form (e.g., potency and related substances assay).

An ideal dissolution test should deliver information in three key areas. First, the dissolution test should be able to detect changes in the physicochemical properties of the drug product from the effect of these changes on the rate or amount of the drug substance released. Such information is useful for the purpose of establishing batch-to-batch production consistency for quality control. Second, dissolution testing should be able to distinguish drug products that

Analytical Method Validation and Instrument Performance Verification, Edited by Chung Chow Chan, Herman Lam, Y. C. Lee, and Xue-Ming Zhang
ISBN 0-471-25953-5 Copyright © 2004 John Wiley & Sons, Inc.

have been manufactured using different processes and/or formulations during the development phase. Finally, when in vitro–in vivo correlation is established, dissolution should also reflect release and absorption rates in humans. However, not all drug molecules are able to fulfill all three functions in a dissolution test.

For a noncompendial drug product (e.g., new drug product), dissolution testing using standard compendial methodology should be developed where possible. The compliance requirements for the *European Pharmacopoeia* (EP), the *Japanese Pharmacopoeia* (JP) and the *United States Pharmacopeia* (USP) must be considered during development and validation of a dissolution test method. Although the USP usually requires determination at a single time point for release testing of an immediate release (IR) drug product, data on drug release at multiple time points should be generated during method development for moderately or sparingly soluble drugs to allow for better characterization of the final drug product.

4.2   SCOPE OF CHAPTER

This chapter outlines the general requirements for the validation of a dissolution method in pharmaceutical products. The development and validation phases of the dissolution test are not as clearcut as in the other drug product testing procedures. Therefore, additional comments on development investigation are sometimes made in this chapter. The discussions are based on method validation for small-molecule pharmaceutical products of synthetic origin and focuses on current regulatory requirements in the pharmaceutical industry. Since expectations for method validation vary at different stages of the product development process, the information given in this chapter is most applicable when validating the final method according to the International Conference on Harmonization (ICH) requirements when preparing for a regulatory submission (e.g., NDA).

Dissolution testing involves a two-step process: sample preparation and sample analysis. In this chapter *sample preparation* denotes the actual sample dissolution procedure, including sample collection. The samples collected from the dissolution apparatus may be analyzed directly or may be subject to further manipulation (e.g., dilution) to give the final sample solutions.

Solid oral dosage forms containing new chemical entities (NCEs) are commonly formulated into tablets or capsules as their first market image formulation. Subsequent drug product line extension development on these NCEs may evaluate more specialized drug delivery systems. Dissolution testing of standard oral tablets or capsules will commonly utilize the paddle or basket apparatus. In this chapter we focus primarily on the development and subsequent validation of dissolution testing methods that use these two devices.

4.3 STRATEGIES, VALIDATION PRACTICES, AND PARAMETERS

The validation requirements are discussed as they apply to both the sample preparation and sample analysis aspects of a dissolution method. The focus of the discussion in this chapter is on the validation considerations that are unique to a dissolution method. *Validation* is the assessment of the performance of a defined test method. The result of any successful validation exercise is a comprehensive set of data that will support the suitability of the test method for its intended use. To this end, execution of a validation exercise without a clearly defined plan can lead to many difficulties, including an incomplete or flawed set of validation data. Planning for the validation exercise must include the following: determination of what performance characteristics to assess (i.e., strategy), how to assess each characteristic (i.e., experimental), and what minimum standard of performance is expected (i.e., criteria). The preparation of a validation protocol is highly recommended to clearly define the experiments and associated criteria. Validation of a test method must include experiments to assess both the sample preparation (i.e., sample dissolution) and the sample analysis. ICH Q2A [1] provides guidance for the validation characteristics of the dissolution test and is summarized in Table 4.1.

The validation requirements are similar to those applied when validating a potency method. Although not listed in Table 4.1, robustness of the various method parameters should be assessed (e.g., the stability of sample solutions). Details of the principles behind these requirements are presented in Section 2.4.

4.3.1 Overview of the Sample Preparation Component

Normally, a definite volume (usually 500 to 1000 mL) of the dissolution medium is introduced into a vessel and the temperature is maintained at $37 \pm 0.5°C$. The testing assembly (i.e., basket or paddle) is fitted to the shaft, and the apparatus

Table 4.1. ICH Q2A Validation Requirements

Characteristic	Requirement[a]
Accuracy	+
Precision	
Repeatability	+
Intermediate precision[b]	+
Specificity	+
Detection limit	−
Quantitation limit	−
Linearity	+
Range	+

[a] +, Signifies that this characteristic is normally evaluated;
−, signifies that this characteristic is not normally evaluated.
[b] In cases where reproducibility has been performed, intermediate precision is not needed.

is adjusted to rotate at a specified rate. The testing assembly is fixed to the shaft following the compendial requirements for each relative position. During operation of the dissolution apparatus, the vessels should be covered appropriately to prevent evaporation of the dissolution medium.

When the basket apparatus is used, the sample is placed in the dry basket; the basket fitted to the coupling disk and lowered to the position specified, and rotation is started immediately. When the paddle apparatus is used, the sample is allowed to sink to the bottom of the vessel, and rotation of the paddle started immediately at the speed specified. If use of the sinker is required, the sample is placed in the sinker and allowed to sink to the bottom of the vessel. The samples are collected at the appropriate time, filtered by a suitable method, and the filtrate used as the sample solution. The drug substance(s) in the sample solution is assayed, and the quantity dissolved at the specified time is expressed as a percentage of the labeled amount.

The requirements for the basket and paddle apparatus described by the three major pharmacopoeias is generally similar but do have some unique differences. These general requirements are summarized in Table 4.2. It is important to know these differences at the time of method development and dissolution. Some of these characteristics are utilized as a system check in the regular performance verification of the dissolution apparatus (e.g., shaft position, shaft rotation variation, and distance of bottom of apparatus to inside bottom of vessel).

4.3.2 Qualitative Dissolution Testing

The visual observation of the dissolution of a dosage form can quickly provide an indication of problems with the formulation or the dissolution test conditions without the requirement for sample analysis. This is particularly useful in the early stages of formulation and method development, when a variety of formulations or a range of dissolution media may be under consideration.

Qualitative assessment of dissolution testing during the initial stage of method development can save a great deal of time, because if certain gross requirements of the test are not met, the sample analysis portion of the test can be eliminated. Examples of some of the possible observations of dosage-form performance and their related issues are:

- The time required for the capsule shell or tablet coating to start breaking apart will provide an indication of possible issues with the shell or coating that would retard the release of the formulation (e.g., gelatin cross-linking).
- The time required for complete disintegration will provide an indication of possible issues with the dosage unit that could affect the release of active ingredient (e.g., overcompression of capsule powders or tablet cores).
- Behavior of the capsule within a specific sinker device (e.g., capsules being stuck to the basket mesh).
- The effectiveness of the mixing within the vessel. *Coning* (the formation of a mound of insoluble excipient particles on the bottom of the vessel) may indicate the need for a higher rotation speed or the use of different apparatus (e.g., baskets instead of paddles).

Table 4.2. General Compendial Requirements for the Basket and Paddle Apparatus

Characteristic	USP	EP	JP
Dissolution vessel	Nominal capacity = 1, 2, or 4 L	Nominal capacity of 1 L	Nominal capacity of 1 L
Shaft position	Not more than 2 mm from the vertical axis	Not more than 2 mm from the vertical axis	Not more than 2 mm from the vertical axis
Allowable variation in shaft rotation speed	±4%	±4%	±4%
Distance of bottom of apparatus to inside bottom of vessel	25 ± 2 mm	25 ± 2 mm	25 ± 2 mm
Apparatus suitability test	Dissolution calibrator, disintegrating, and nondisintegrating types	Not specified	Not specified
Temperature of dissolution medium	37 ± 0.5°C	37 ± 0.5°C	37 ± 0.5°C
Additives in dissolution medium	Purified pepsin up to an activity of 750,000 units/1000 mL or pancreatin up to an activity of 10 USP units/1000 mL	Not specified	Polysorbate 80 up to 1.0% w/v[a]
Sample withdrawal	Midway between surface of medium and top of rotating basket or blade; not less than 1 cm from the vessel wall	Midway between surface of medium and top of rotating basket or blade; not less than 1 cm from the vessel wall	Midway between surface of medium and top of rotating basket or blade; not less than 1 cm from the vessel wall

(continued overleaf)

Table 4.2 (*continued*)

Characteristic	USP	EP	JP
Allowable dosage unit sinkers	Wire helix or other validated sinker devices	Suitable sinker (e.g., wire or glass helix)	Sinker with defined format[b]
Data interpretation	6 + 6 + 12	6	6 + 6
	S1 6 Each unit is not less than $Q + 5\%$	All 6 units are not less than Q	All of the first 6 samples or 10 out of 12 meet specified requirements
	S2 6 Average of 12 units $(S1 + S2)$ is equal to or greater than Q, and no unit is less than $Q - 15\%$		
	S3 12 Average of 24 units $(S1 + S2 + S3)$ is equal to or greater than Q, not more than 2 units are less than $Q - 15\%$, and no unit is less than $Q - 25\%$		

Source: Refs. [2]–[4].

[a]Ref. [5].

[b]12.0 ± 0.2 mm in inside diameter and 25 to 26 mm in length. The trunk consists of a spiral 3.0 to 3.5-mm pitch coil made with an acid-resistant 1-mm-thick wire. The spiral is supported on the outside with 10 wires which are fixed almost in parallel and with an equal distance. The sides are fixed with two double wires in a cross shape, but on one of the sides, it is fixed with a clasp and can be opened so that a sample can be inserted.

- The suitability of the media degassing method. The formation of bubbles during the dissolution process can affect the rate of release of the active ingredient.

Table 4.3 shows the results of dissolution of a capsule formulation. Two sinkers were investigated in a series of experiments. Qualitative assessment of dissolution experiments was conducted to compare the behavior of the capsule when different sinkers were used. These experiments culminated in selection of the preferred sinker (i.e., sinker B) for this formulation. Once the analysis was repeated using the preferred sinker device, the dissolution test showed good results with a low degree of variability.

4.3.3 Validation of the Sample Preparation Component

Different approaches may be used to validate the sample preparation component of the dissolution test. However, it is important to understand that the objective of validation is to demonstrate that the procedure is suitable for its intended purpose. For example, one of the strategies will demonstrate the validity of different aspects of sample preparation during method development (prior to the formal method validation exercise). As a result, the final validation experiments will confirm the work done during method development. The strategy that will be followed for the method development and validation process will depend on the culture, expertise, and strategy of the analytical laboratory.

Apparatus. The nature of the dosage form will determine the type of dissolution apparatus that will be used for method development and validation. The following questions must be asked when selecting the dissolution apparatus. Is it a capsule? Will a sinker be required? How stable is the drug substance after dissolution in the medium? Is the formulation an immediate release or an extended release formulation? Is this a transdermal patch?

Table 4.3. Physical Observation and Analytical Variability at 15 min of Dissolution

	Type A Capsule Sinker		Type B Capsule Sinker	
Run	Observation	% RSD of Amount Released	Observation	% RSD of Amount Released
1	Normal disintegration	1.6	Normal disintegration	3
2	Normal disintegration	1.7	Normal disintegration	1.7
3	Some gelatin cross-linking (membrane formation)	13.9	Normal disintegration	1.3
4	Some gelatin cross-linking (membrane formation)	32.1	Normal disintegration	1.7

USP Dissolution Apparatus 1 (basket) and 2 (paddle) are commonly used for immediate-release formulations. USP Apparatus 3 (reciprocating cylinders) is the system of choice for testing extended-release products or a dosage form that requires release profiling at multiple pH levels and time points. Low-dose products may require the use of flow-through analysis or other low-volume test techniques (noncompendial 100- or 200-mL dissolution vessels). Once the apparatus is selected and has been shown to be suitable during method development, no further evaluation of another apparatus is required during validation.

Dissolution Medium. Water, hydrochloric acid (0.1 N), and various pH buffers are commonly used dissolution media. Although commonly used as a dissolution medium, water provides no control of pH and should be avoided. The pH of the water can be influenced significantly by the formulation components (including the active ingredient). Lack of pH control can result in a change in the release profile. A pH change may be the result of normal variability in excipients, or changes in the formulation components as a result of degradation. Hydrochloric acid (0.1 N) is often used as a dissolution medium because it mimics the acid environment of the stomach. Other dissolution media (e.g., pH 4.5 or 6.8 buffers) can be used to mimic the expected patient gastric state (i.e., fasted or fed) or to improve the release profile and/or discrimination. In the case of low-solubility compounds, a surfactant (i.e., polysorbate 80) may be utilized to improve the dissolution profile.

The dissolution medium that is chosen for method development and validation will depend on a variety of factors that include:

- The solubility of the drug molecule
- The nature of the dosage form
- The chemical structure of the drug molecule

Deaeration is a very important consideration in the development and validation of a dissolution test, as it can affect the rate of release of the drug substance from the dosage form. Ideally, a method should not be affected by the deaeration procedure. At a minimum, it should be demonstrated that some variability in the degree of deaeration will not significantly alter the results of the dissolution test. It should also be noted that media containing surfactants should not be deaerated, as that can result in excessive foaming.

There are three common ways to deaerate a dissolution medium: (1) vacuum filtration, (2) helium sparging, and (3) heating. Vacuum is commonly applied after filtration of the dissolution medium, with continued exposure of the filtrate to the low vacuum created by the aspirator (with or without heating). Potentially, the water pressure (i.e., degree of vacuum) of the water aspirator can affect this method of deaeration. Care should be taken to ensure that adequate suction has been applied. The time of exposure should be noted.

Helium sparging is commonly used to displace the dissolved air in HPLC mobile phase. The same principle can be used for deaeration of media. The time of helium sparging should be noted, as this can be a critical parameter in the dissolution test.

Heating is the least common of the three techniques used to deaerate the dissolution medium. In this technique, the filtered medium is heated above 37°C (up to about 90°C) with stirring to drive off the dissolved air. The temperature and time interval used is critical to determine the extent of the deaeration process.

An indication of the effectiveness of the deaeration technique selected can be obtained by measurement of the final oxygen level in the medium. Deaeration should occur immediately prior to the use of the medium to prevent reaeration. However, to perform deaeration just prior to the point of use is not always practical. Therefore, one should generate data that support the use of media that have undergone some level of reaeration and still show acceptable levels of dissolved oxygen.

Rotational Speed. The rotational speed of a basket or paddle is an important consideration in the development and validation of the dissolution test. A speed of 100 rpm is commonly used with the basket apparatus and a speed of 50 rpm is used with paddles. In method validation, one needs to ensure that slight variations in rotational speed will not affect the outcome of the dissolution test. The compendial limit for variations in rotational speed is ±4%, but a wider variation (e.g., ±10%) may be considered in testing the robustness of the method.

Sample Collection. The two aspects of sample collection that need to be considered during the development and validation of sample preparation are (1) the withdrawal of the sample aliquot from the dissolution vessel and (2) the clarification (filtration) of the sample aliquot. The availability of an automated versus a manual sample collection set up in the quality control laboratory needs to be considered when developing and validating the method. If automated sample collection is chosen, equivalency must be demonstrated compared to the manual method of sample collection. The potential effect of carryover from the tubing in the automated sampling system may give rise to a positive bias. This must be investigated to ensure that if any carryover occurs, it falls within acceptable limits. Depending on the amount of carryover, it may become necessary to implement a specific cleaning process for the system to ensure that carryover is minimized. On the other hand, the effect of adsorption in tubing will cause a negative bias. If the bias is too high, it may be necessary to restrict sample collection to a manual method. Finally, the sampling probe has the potential to alter the hydrodynamic flow within the vessel. This needs to be considered when comparing automated and manual sampling. Ideally, the sampling probe should be immersed in the vessel at the time of sample collection only.

Filters are required for dissolution sample collection. It is necessary to filter out the excipents that may cause interference in sample analysis. Appropriate recovery studies should be performed and documented. Any observed bias should be addressed. Filtration must be performed when the sample aliquots are withdrawn, not at a later time.

Effect of Non-USP Adaptation. The development and validation of the dissolution test for new formulations will often probe into the use of nonpharmacopoeia adaptations (e.g., peak vessel and unique sinker configurations). The suitability of these adaptations must be assessed during method development or validation.

Cleaning Validation. Once the vessels have been cleaned, a "blank" dissolution run needs to be performed to ensure that the cleaning procedure for the dissolution vessels is appropriate and will not give rise to contamination. Any unique cleaning steps must be identified during development or validation of the method and included in the test procedure. The robustness of the sample collection procedure can be studied in an experimental design that will investigate all or some of the parameters discussed previously. Table 4.4 shows the summary data of a statistical analysis of a fractional-factorial experimental design of 44 runs. The experiment was designed to study the effect of deaeration, media concentration, paddle height, paddle rotation, and sampling time. In this manner, the normal variations in the method operating conditions were simulated.

The p-value for media concentration, paddle height, and sinker-type factors indicated no statistical significance (p-value > 0.05). However, even though statistical significance was observed for media deaeration, paddle rotation, and sampling time, the contribution of these effects was estimated to be insignificant.

4.3.4   Validation of the Analytical Component

As mentioned earlier in the chapter, validation of the analytical component of the dissolution test will follow guidelines similar to those described in Chapter 2, where the validation parameters are discussed in detail. For the purposes of this chapter, only an overview is provided, with emphasis on the unique requirements of the dissolution test.

Linearity. Neat standard solutions are prepared to cover the nominal concentration of the sample. ICH Q2B [1] proposes an acceptable range of $\pm 20\%$. A range of 25 to 125% of nominal concentration is commonly used for the linearity determination. This range allows quantitation at the early dissolution time point.

Table 4.4. Summary of JMP Robustness Analysis

Factor	Factor Range	p-Value	Effect Estimate (% Dissolution)
Deaeration	Yes/no	0.0059	0.5
Media concentration	0.08–0.12 N	>0.05	0.1
Paddle rotation	45–55 rpm	0.0002	0.9
Paddle height	15–35 mm	>0.05	0.1
Sinker type	3-prong/spiral	>0.05	0.3
Sampling time	13–17 min	0.0014	0.7

Visual inspection of a plot of response versus concentration will show a straight line. The correlation coefficient (r), residual sum of squares, and y-intercept should be reported. In the case of the dissolution profile for a modified release product, ±20% of the stated range will need to be prepared. For example, a range of 0 to 110% of release will be required for a profile of 20 to 90% release.

Accuracy. Sample solutions of known concentration (e.g., spiked placebo) are used for the accuracy determination. Experimental work may be organized so that the same stock solutions are used to prepare both linearity and accuracy solutions. The accuracy solution must be exposed to normal test conditions (e.g., mixing in a heated dissolution vessel). Determine any bias that is caused by the sampling and analysis of the solutions. If a dissolution profile of the drug product is required, accuracy determinations at different concentrations of the required profile will need to be performed (e.g., at 40, 75, and 110% of theoretical release). The results are reported as percent theory.

Precision: Repeatability. The repeatability experiment will be performed using six dissolution samples that are prepared from the same dissolution apparatus.

Precision: Intermediate Precision. Sets of six dissolution samples that are prepared using different instruments and by different analysts are used to determine intermediate precision. However, this procedure will not be able to differentiate method variation versus tablet-to-tablet variation. It will predict the worst-case precision that includes tablet-to-tablet, sampling, and analysis variations.

To determine precision at multiple collection time points for the dissolution profile of a modified release formulation, normalization to the final time point (or infinite time point) will eliminate tablet-to-tablet and lot-to-lot variation. Figure 4.1 illustrates the way that normalization is used to remove tablet-to-tablet variation. However, the normalization technique should be used during development only as a means of investigation of the profile. The final formulation should be sufficiently robust to produce complete release at its final time point. Normalization to remove lot-to-lot variation can be performed using the following equation:

$$\%_{\text{normalized}} = \frac{\%_t}{\text{lot potency}}$$

where $\%_t$ is the percent dissolved at time t.

Range. The results of the linearity, accuracy, and precision will help determine the range for the dissolution test (e.g., 25 to 125% of nominal for a single-point dissolution and ±20% of the stated range for a dissolution profile for a modified release formulation).

Robustness for HPLC Analysis. The investigation of the effect of column, mobile phase, HPLC solution stability, and wavelength is performed in a manner similar to the HPLC potency/related substance assay. For solution stability, the

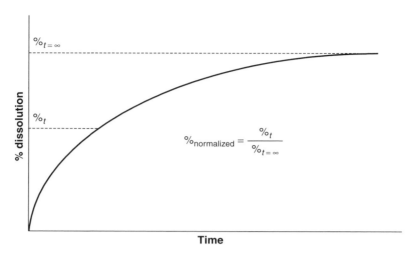

Figure 4.1. Normalization of dissolution profile to remove tablet-to-tablet variation.

solution can be analyzed on different days or analyzed on the same day with both aged and fresh solutions.

Robustness for UV–Vis Analysis. Wavelength accuracy, wavelength repeatability, diluting solvent (i.e., pH, concentration), solution stability, and bubble formation by the sipper can be investigated during validation of the analytical component.

Specificity. For HPLC analysis, resolution of the drug substance from any potential excipient and system interference peaks should be demonstrated. For a UV–Vis analysis, the absorption of the placebo solution should not be significant. It is important to note that the dissolution test is not intended to be stability indicating and will not need to be able to separate degradation peaks from the analyte peak.

4.4   DISSOLUTION METHOD REVALIDATION

There are various situations during the life cycle of a dissolution test that will require revalidation of the method. These are similar to those described for the potency assay in Chapter 2.

4.5   COMMON PROBLEMS AND SOLUTIONS

In the following pages we summarize the common deficiencies of a dissolution method which may result in problems during the validation process. The common

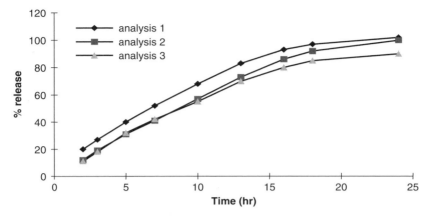

Figure 4.2. Negative bias of dissolution profile.

problems for the analytical component are similar to those described for the potency method in Chapter 2.

1. *Negative bias of the dissolution test.* Figure 4.2 illustrates three analyses, where analyses 2 and 3 are consistently biased low compared to analysis 1. Analysis 1 represents the normal curve describing 100% release of the drug substance.

Several potential areas can cause negative bias during method validation:

- The impact of the standard curve and analyte response linearity.
- Larger negative bias noted for lower sample concentrations due to analyte binding to a variety of materials, including excipient, apparatus surfaces, and/or filters.
- Larger negative bias noted for higher sample concentrations due to poor sample solubility, resulting in sample segregation (precipitation) after collection as the solution temperature drops from 37°C to ambient (or refrigerated) conditions for analysis.
- Other consistent negative bias regardless of sample concentration:
  - Sample solution composition does not match the standard solution, resulting in a low response bias for samples. This can be due to a standard/sample preparation scheme or to negative formulation matrix effects in the dissolution media (e.g., pH change).
  - Sample degradation during and after the dissolution process can change sample solution response compared to that of standards.
  - Sample calculations for multipoint (profile) analyses that do not correct for sample and media volume removed at earlier time points can bias results. The bias increases with increasing sampling volume and increasing number of sampling time points.

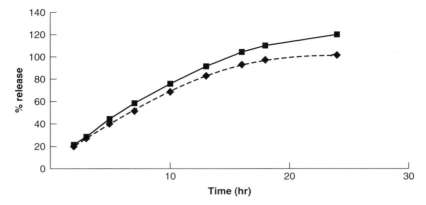

Figure 4.3. Positive bias of dissolution profile.

2. *Positive bias of the dissolution test.* Figure 4.3 shows positive bias for an analysis that is above the normal curve, which gives 100% release.

Potential causes for positive bias are:

- The impact of the standard curve and analyte response linearity.
- Larger positive bias noted for lower sample concentrations, due to positive interference from equipment or materials such as excipients, vessel residue, sampling equipment, and filters.
- Consistent positive bias regardless of sample concentration:
  - Sample degradation during and after the dissolution process can change the sample solution response compared to standards depending on the absorptivity of the degradant if direct UV–Vis analysis is used.
  - Sample solution composition does not match the standard solution, resulting in a high response bias for samples. Can be due to a standard/sample preparation scheme or to positive formulation matrix effects in the dissolution media (e.g., pH change).
  - Evaporative loss can lead to biased results, particularly for extended dissolution periods (e.g., several hours to several days).

3. *Calibration of dissolution equipment.* It is very important to ensure that the calibration of the dissolution equipment is kept up to date. The expiry date of the calibration status should be checked each time that dissolution testing is performed.

4.6 SUMMARY OF DISSOLUTION VALIDATION DATA

It is very useful to summarize all method validation data into a tabular format. The tabulated summary will give a quick overview of the validation data. The tabular format should give details of the ICH requirement for validation as well as the results achieved. In conclusion, the necessary data have been generated in support of the method validation. Table 4.5 gives an example of how the data can be recorded.

Table 4.5. Validation Summary

ICH Validation Characteristic	ICH Recommended Requirement	Validation Results
Accuracy	Assessed using a minimum of nine determinations over a minimum of three concentration levels covering the range specified. Report as % recovery.	Avg. % recovery = 100.39%; % RSD = 1.8% ($n = 9$).
Precision		
Repeatability	The worst-case relative standard deviation (RSD), for at least six replicates for a range of doses.	The RSD of individual results for 15-, 30-, and 45-min sampling times ranged from 2.1 to 3.1% RSD ($n = 36$).
Intermediate precision		The % RSD of the average dissolution results (of $n = 6$ dosage units) after 15-, 30-, and 45-min sampling times ranged from 1.6 to 2.6% ($n = 6$ average results for each time).
Reproducibility		The average dissolution results obtained at 15-, 30-, and 45-min were within 2 to 8% dissolved between laboratories.
Specificity	Representative chromatograms to demonstrate specificity.	No interference was observed from media or excipients. A reference chromatogram is attached.
Linearity	Data from regression line (correlation coefficient, y-intercept, slope, residual sum of squares) and a plot.	Correlation coefficient = 1.0000; y-intercept = 0.2040 area unit; slope = 38.1669 area units (μg/mL); residual sum of squares = 84.6069. A plot of linearity data is provided.

(*continued overleaf*)

Table 4.5 (*continued*)

ICH Validation Characteristic	ICH Recommended Requirement	Validation Results
Range	Confirm procedure provides an acceptable degree of linearity, accuracy, and precision when applied to samples containing analyte within or at extremes of the range of procedure specified.	Confirmed within the range 25 to 125% of test concentration based on linearity and accuracy data.
Robustness	Critical factors include paddle height and rotation, media concentration, sinker type, media deaeration, and sampling time.	Of the critical factors evaluated, media concentration, paddle rotation, and sampling time indicated a statistically significant effect ($p < 0.05$). None were found to be significant. Working standard and sample solutions were assessed to be stable for 4 days.

REFERENCES

1. ICH Harmonized Tripartite Guidelines, ICH Q2A, *Text on Validation of Analytical Procedures*, Mar. 1995; ICH Q2B, *Validation of Analytical Procedures: Methodology*, May 1997.
2. *European Pharmacopoeia*, 4th ed., Section 2.93, Dissolution Test for Solid Dosage Forms, 2002.
3. *United States Pharmacopeia*, USP 26 Chapter ⟨711⟩, Dissolution, 2003.
4. *Japanese Pharmacopoeia*, 14th ed., Chapter 15, Dissolution Test, pp. 33–36, 2001.
5. *Japan Ministry of Health & Labour Guidelines*, PAB/PCD No. 487, Dec. 1997.

5

DEVELOPMENT AND VALIDATION OF AUTOMATED METHODS

CHANTAL INCLEDON AND HERMAN LAM, PH.D.
GlaxoSmithKline Canada, Inc.

5.1 INTRODUCTION

The use of automated sample preparation equipment in the laboratory is becoming more common as the pace of technology in the pharmaceutical industry continues to increase. The intensifying pressure to reduce the time line and cost of drug discovery and development has caused the pharmaceutical industry to turn toward automation. Laboratories that once viewed automation as an expensive luxury are now looking to automation as a solution to increase sample throughput, help ensure data integrity, and improve on laboratory safety [1]. Unfortunately, guidance on the development and validation of automated methods is limited. The purpose of this chapter is to provide guidance and operational tips to industry for the development and validation of automated sample preparation procedures that meet current Good Manufacturing Practice requirements.

5.2 SCOPE OF CHAPTER

The contents of this chapter are applicable to both product development and QC laboratories. Although the principles discussed in the chapter are presented in the context of immediate-release solid dosage forms, they can be applied to modified-release and suspension dosage forms if it is within the capability of the

Analytical Method Validation and Instrument Performance Verification, Edited by Chung Chow Chan, Herman Lam, Y. C. Lee, and Xue-Ming Zhang
ISBN 0-471-25953-5 Copyright © 2004 John Wiley & Sons, Inc.

equipment. Guidance is provided on the following analyses: (1) multiple- and single-dosage-form assays, (2) content uniformity, (3) degradation products and synthetic impurities, and (4) dissolution testing. Any analytical method can be viewed as being composed of the following elements:

$$\text{sample preparation} \rightarrow \text{measurement} \rightarrow \text{data handling}$$

The information in this chapter applies specifically to the first element: sample preparation. The sample preparation steps are usually the most tedious and labor-intensive part of an analysis. By automating the sample preparation, a significant improvement in efficiency can be achieved. It is important to make sure that (1) suitable instrument qualification has been concluded successfully before initiation of automated sample preparation validation [2], (2) the operational reliability of the automated workstation is acceptable, (3) the analyte measurement procedure has been optimized (e.g., LC run conditions), and (4) appropriate training in use of the instrument has been provided to the operator(s). The equipment used to perform automated sample preparation can be purchased as off-the-shelf units that are precustomized, or it can be built by the laboratory in conjunction with a vendor (custom-designed system). Off-the-shelf workstations for fully automated dissolution testing, automated assay, and content uniformity testing are available from a variety of suppliers, such as Zymark (*www.zymark.com*) and Sotax (*www.sotax.com*). These workstations are very well represented in the pharmaceutical industry and are all based on the same functional requirements and basic principles.

When developing and validating an automated procedure, the existence or absence of a validated manual method will determine the subsequent course of action taken. If a manual method exists, the automated method can be developed based on the manual method. Validation, in broad terms, will consist of demonstrating equivalence between the automated and manual methods. If a manual method does not exist, the automated method can be developed and validated without developing a manual method first. Validation will place more importance on demonstrating the accuracy and robustness of the automated method since equivalency testing is not applicable. This approach does require consideration of the fact that regulatory agencies may require samples for testing within their laboratory using a registered manual method, or they may have to purchase and assemble the automated equipment used in the validation. This will, of course, affect the approval time lines. Another consideration is that the methodology within a company may need to be transferred to a laboratory that does not have automated equipment.

When creating an automated sample preparation to mimic or reproduce a manual sample preparation procedure, it is wise to break down each step of the procedure and understand its relevance. When transferring the manual steps to automated steps, not every step must or can be transferred individually. For example, when developing a method there might be volume restrictions due to the physical capability of the instrument components. If the manual method requires the addition of 25 mL of dissolving solvent to the tablets as the first step, followed by sonication and then making up to volume with solvent, the automated sample preparation may not be able to mimic each step exactly. The Zymark Tablet Processing Workstation (TPW II) is designed such that a minimum volume addition of 50 mL is required prior to the homogenization step to disperse the tablets. Therefore, the automated sample preparation will need to be modified slightly to accommodate the limitations of the system.

Manual Method	Automated Method
Add 25 mL of solvent to tablets.	Add 50 mL of solvent to tablets.
↓	↓
Sonicate for 30 minutes.	Homogenize for 15 minutes.
↓	↓
Make up to volume to 100 mL with solvent.	Add approximately 50 mL of solvent.

Final volume using the automated method will be 100 mL minus tablets volume displacement (refer to Section 5.3.1).

Usually, the tablet dispersion method used in the automated method will be different from the dispersion method used in the manual method. It cannot be assumed that the same amount of time will be required to disperse the tablets by two different means. If the automated method uses homogenization instead of 30 min of sonication, it is necessary to determine the amount of homogenization time required. Therefore, the first step in automating a sample preparation when a manual procedure exists is to examine the manual procedure carefully, understanding the reason for each step and evaluating if the manual method is automation "friendly." Some important considerations when converting a manual method to an automated method are maintaining the same column loading and replacing multiple manual steps (such as two-step dilutions) by one automated step. Automated workstations usually employ sequential processing versus the batch processing generally used in manual methods. To maximize the efficiency of the automated procedure, when using online HPLC analysis the sample preparation time should coincide with or be shorter than the HPLC run time.

Creating an automated procedure for sample preparation involves two stages, method development and method validation [3–6]. The purpose of the method development step is to assess the parameters of the method critically to ensure that the procedure is suitable for the intended sample with acceptable accuracy and precision. The parameters that are typically assessed during method development are discussed next. Some of these parameters may also apply to the manual method.

5.3 DEVELOPMENT AND VALIDATION PRACTICES

5.3.1 Method Development

Solution Density Determination. Some automated systems handle solution dispensing gravimetrically by taring the vessel, dispensing the solution, and then weighing the vessel again with a three-or four-decimal-place balance. The volume dispensed is calculated by converting the weight to volume using the solution density value. For systems with this capability, knowledge of solvent density used on the workstation becomes essential. Density can be determined using a densitometer or by weight of a known volume of solution. The density determination should be measured for all diluting solvents and filtered extracted sample aliquots if a second, or two-step, dilution is incorporated in the automated procedure. In the case of dissolution, the solvent or medium, is often dispensed into the vessels at a temperature of 37°C. Therefore, the density must be determined with the solution at this temperature to ensure accurate dispensing by the automated workstation. The density value can be incorporated as a fixed value in the validated method when sufficient confidence in the value has been established. A density check can be added to the method to measure the density of the solvent(s) prior to the analysis to confirm accurate density within a fixed range for use. A density range of ±0.5% of the established value is an acceptable range for density confirmation.

Probe Hold-up Volume. Automated systems such as the TPW II make use of a probe to dispense solutions. When working with a low volume, for example during content uniformity sample preparation, it may be necessary to determine the volume of solution retained in the probe. The probe is immersed in the solution for homogenization, and when the probe is withdrawn from the solution, a small volume of solution is retained inside the probe and on its surface. After the first sample is processed, the second sample will be exposed to the weight of the solvent actually added to the vessel as well as the weight of the solvent that remained on and in the homogenization probe. The weight of solvent that remained on the probe, the *probe hold-up volume*, should be determined. It is advisable to avoid the need to test this parameter by using large volumes where the fluctuation in probe hold-up volume is not significant to the target volume of the solvent for extraction. Some workstations, such as the Zymark TPW II, have a feature that can be activated to automatically determine the probe hold-up volume at the beginning of the run. Typical probe hold-up volumes for a 250-mL dispenser range between 0.2 and 0.6 mL.

Volume Displacement. This parameter is not a factor in dissolution testing but can prove to be a very important factor in automated assay, content uniformity, or degradation and impurities testing. It specifically addresses the volume displaced by the tablets in solution. Since manual sample preparations are often prepared utilizing volumetric flasks where the solution is diluted to the mark, the actual volume of solvent added to the flask is irrelevant. However, this actual volume of solvent added becomes very important when comparing a volumetric manual method against a gravimetric automated method. If the displacement volume is significant, the results of the manual and automated methods will differ. For the automated method, the volume of solvent added to the tablets must be stated. Since the final concentration of the sample must be the same in the manual and automated methods, the actual volume of solvent added to the flask must be determined. In fact, even if a manual method does not exist, it is recommended that the volume displacement be calculated in case the need for a manual method arises at a later date. It would prove difficult to develop an equivalent manual method if the volume of solvent to be added to the sample could not be achieved by the use of a volumetric flask. The actual volume of solvent added for both the manual and automated methods will be the final volume of the solution (sample + solvent) minus the volume displaced by the sample. An example for the determination is as follows. Five tablets are added to a 250-mL volumetric flask and weighed (weight = 484.44 mg). Some solvent is added and the sample is shaken to allow the tablets to disintegrate. Once the solution is at room temperature, the sample is diluted to the mark with solvent. The volumetric flask is weighed again (weight = 730.82 g). Taking into account the density of the sample solution (density = 0.9965 g/mL), the actual volume of solvent added is determined (730.82 g − 484.44 g = 246.38 g/0.9965 g/mL = 247.25 mL). This value (247.25 mL) is then entered into the automated method instead of 250 mL.

System Compatibility. Testing for system compatibility should be performed at an early stage of method development to ensure that all parts of the system that are in contact with the sample or solvents are compatible with the product and all solvents used. Unlike the manual method, the automated method sometimes relies on an intrusive extraction mechanism whereby the sample solutions are exposed to the extraction source as in the case of homogenization. It is therefore imperative that the extraction solvent, drug, and excipients do not interfere by extracting material from the system or adsorb to any part of the system. Close attention should be paid to system compatibility, since adsorption of the active ingredient to any part of the system could prove to be a very challenging obstacle to the success of the automated method. If acidic extraction media are used, damage to the components of the automated system can result and the necessary precautions, such as alternative construction materials for bushings in the homogenizer probe or tubing, may be required. To establish whether there is interaction between the system and the active ingredient or the solvent, an automated sample can be prepared using an empty tube and a standard solution as solvent. The prepared sample is then analyzed and compared to a manual preparation of the standard. Once it is established that there are no interactions, verification that there are no interactions between the system and the excipients should be completed. A placebo sample should be prepared using the automated workstation and the response of any observed system or placebo peaks should not be greater than 1.0% of that for the standard solution. For degradation and impurity methods, any additional peaks ≥ limit of quantitation (LOQ) should not interfere with the active component, any known synthetic impurities, or any degradation products.

Temperature. During extraction of the sample, the homogenization process may cause a significant increase in the temperature of the extraction solvent. This can lead to dilution errors caused by changes in solution density. Whenever possible it is best to avoid this scenario by carefully designing the extraction steps of the automated method and by choosing solvents and a process that will minimize heating of the solvent. Usually, the process of performing numerous short pulses instead of fewer pulses of longer duration will tend to reduce the heating of the solution. For the extraction process, the solution should be programmed to stand for a certain period of time to allow it to cool to room temperature before proceeding with the sample preparation if significant heating is observed. If solvent heating is unavoidable, it may be possible to correct for the change in density in the calculation once the correlation between temperature and density is understood.

Extraction. Determination of the efficiency at which the analyte is extracted is critical to the development of a successful automated method for assay, content uniformity, or degradation and impurities. Critical parameters such as extraction time and speed should be investigated and should be varied across a suitable range to determine the extraction profile. The extraction conditions should be chosen from the plateau of the extraction profile to ensure ruggedness of the extraction method. Many companies make use of experimental design to map the extraction area, by varying the number of pulses, the duration of the pulses, the speed of the pulses, and the insertion of soak/settle times. The extraction study can be the most time-consuming activity of the method development process but is invaluable to the flexibility of application of the automated method. A well-designed and executed extraction study will ensure the robustness of the automated method, such that small operating differences between automated platforms will not affect the results of the automated method. Furthermore, physical changes in the product upon storage during a stability study should not render the automated method useless. Consider testing product stored under stressed conditions of elevated temperature and humidity during the method development phase of the automated method.

Filtration. Suitability studies should be conducted to ensure that analyte is not retained on the filter membrane to any significant extent. The filtration study can be broken down into four components: determination of prewet volume, determination of filter capacity, active absorption profile, and extractable profile. Determination of the extent of interference present in the filtrate from particulate or membrane extractables is especially important for spectrophotometric assays. The filtration study should aim to assess the drug retention profile for the filter by determining the waste volume beyond which the filter profile showed no significant retention of active sample, as well as the point in the retention profile where the analyte is being retained again due to filter blockage. The filter capacity is an important consideration because it can lead to failure of the automated system if the formulation excipients block the filter. This is more prominent for automated testing of a multipoint dissolution profile where it would be more cost-effective to use one filter per vessel for all sampling time points of the dissolution run. Considering the potential blockage, the safest route is to change the filters at each time point. However in the case of many consecutive dissolution runs with multiple sampling time points, this procedure could easily add up to hundreds of filters.

The filter study can be performed manually off-line, but it is preferable to conduct it on the automated system since the pressure loading on the filter will differ between manually filtered samples and samples filtered by the automated system. Two parameters must be defined: prewet volume, equivalent to the volume filtered to waste during a manual filtration, and collect volume. To determine a suitable filter prewet volume, filter and analyze a solution in 1- or 2-mL increments over a suitable volume range (typically, 10 mL for content uniformity and assay methods and 20 mL for dissolution) to determine the retention profile (Figure 5.1). The filter prewet volume should be selected such that it is well within the plateau of the retention profile graph. This test should be performed on the least concentrated sample solution because it is the sample most likely to yield significant filtration loss. For degradation and impurity methods, the test should be performed on a sample with a typical degradation profile.

Having determined the optimum filtration conditions, filter a sample and compare its response to a centrifuged or unfiltered sample (reference sample). To determine the filter's volume capacity for dissolution testing, filter the sample through the filter in increments equal to the volume used by the automated workstation at each sampling time point until the desired maximum volume is achieved. Compare the response of each filtered aliquot with the response of the centrifuged or unfiltered aliquot. This will determine at which sample time point the filters must be changed and should be performed on the sample solution that contains the highest level of insoluble excipients.

For content uniformity, assay, and dissolution assays, a filtered aliquot is acceptable if the difference between the response of the filtered aliquot is not greater than ±1.0% of the response of the reference sample. For impurities methods, the response of each impurity after filtration should be within ±20% of the

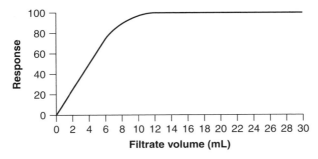

Figure 5.1. Filter retention profile.

reference sample response. It is suggested to use a volume safety factor of 1 mL or greater when setting the prewet volume and volume capacity parameters in the automated method.

Using the optimized filtration conditions, filter a blank sample solution (sample solvent) to determine whether any materials are leached or extracted from the filter's membrane or housing. The response from the filtered sample solvent should not be greater than 1.0% of the average response for the unfiltered standard solution. In addition, for separative finish (e.g., HPLC), no peaks should coelute with the principal peak of the chromatogram.

Dilution. It is important to be aware that the dilution process utilized by the automated workstation may differ from a manual process. Adjustments in dilution factors may be required and considerations of the density value of the diluting solvent and sample filtrate are valuable. In some cases the dilution step can be avoided when designing the automated sample preparation through initial volume adjustment whenever possible and/or through changes in the injection loop size to achieve proper column loading.

Carryover. Carryover studies are performed to ensure that no significant level of the analyte is introduced to a sample from the previous sample. Carryover can be estimated by determining the amount of active present in a blank preparation injected after the active sample. It should also be shown that no cumulative carryover effects are occurring. Although a blank preparation, analyzed after the analysis of one sample, may not show any active carryover, it is possible that after 10 samples, a significant amount of active is being carried over to the subsequent samples. For content uniformity, assays, and degradation and impurities methods, a carryover of less than 1.0% of the active concentration is acceptable. If the carryover observed is higher than the set limit, the vessel wash and the filter and transfer path wash can be increased or modified to resolve the carryover problem. In some cases the solvent used for the washing cycle may need to be changed to another solvent.

For dissolution, both the interrun carryover and the intrarun carryover must be assessed. The interrun carryover, which deals with the carryover from one run to another, can be determined by means of a blank dissolution run that immediately follows an active sample dissolution run. For cumulative carryover, a blank dissolution run would be performed after a series of active sample dissolution runs (minimum of three runs). The intrarun carryover, which deals with the carryover between successive samples in a particular dissolution run, is usually relevant if a common sample pathway is used for all vessels. In this case, a dissolution run with active sample in alternate vessels interspersed with a blank dissolution run (active sample in vessels 1, 3, and 5 and no active sample in vessels 2, 4, and 6) should be tested. A carryover of less than 1.0% of the active concentration in the blank vessels applies. If the carryover observed is higher than the set limit, adjustments to either the flush volume or the vessel wash parameters can be made to eliminate the carryover.

Stability of Solutions. If applicable data are not available under the conditions experienced in the automated workstation, suitable studies should be performed. Solution stability must be assessed for both degradation and evaporation. For degradation, solutions containing active sample and placebo should be prepared and stored in the automated workstation under typical environmental conditions for the length of the desired stability before reassaying. For content uniformity, assays, and dissolution methods, the response of the solution should be 98.0 to 102.0% of the initial value. For degradation and impurity methods, the profiles should be similar, all impurities above LOQ should be present, and the profiles should agree with the following:

Impurity Level	Absolute Agreement
<0.5%	±0.1%
0.5–1.0%	±0.2%
≥ 1.0%	±0.3%

To test for evaporation, the weight loss should be determined on suitable aliquots of a test solution stored in the automated workstation under typical environmental conditions for the length of the desired stability before reassaying. When compared to the initial weight of the test solution, the weight change should not be greater than 2.0% and the response of the solution should be between 98.0 and 102.0% of the initial value.

5.3.2 Method Validation

The level of testing performed at this stage should be appropriate for the intended use of the procedure [7–9]. For instance, it may be acceptable to minimize testing during preclinical and phase I studies. The information provided in this section describes the validation requirement suitable for phase III development or QC activities. The acceptance criteria discussed are for guidance only and should be assessed on a product-by-product basis, depending on the precision and accuracy of the analytical method. Although at the method development stage, it is not required to devise a formal method development protocol, it is advisable to prepare a formal method validation protocol prior to the start of the validation. The validation protocol and all acceptance criteria should be justified scientifically and suitable for the product being analyzed. The following aspects are to be considered when validating the automated procedure: (1) precision, (2) accuracy, (3) equivalency, and (4) robustness. The linearity, limit of quantitation (LOQ), and limit of detection (LOD) are usually addressed through validation of the measurement portion of the analysis. The validation protocol should describe how each of these aspects are going to be tested, how many samples will be prepared and analyzed, and list the agreed-upon acceptance criteria.

Precision. The precision of an analytical procedure expresses the closeness of agreement between a series of measurements from the same homogeneous sample over a period of time under the method conditions prescribed. Precision

Table 5.1. Testing for Repeatability

Method	Testing	Acceptance Criteria
Assay	Six sample determinations at the nominal concentration.	The variability of the automated method should not be more than 2.0% (RSD) or less than the manual method.
Degradation and impurity	Six determinations of a sample spiked at specification limit with available impurities.	The variability of the individual impurities is not more than 15.0% (RSD), and total impurities is not more than 10.0% (RSD) or less than the existing manual method, whichever is greater.
Content uniformity	Ten sample determinations at the nominal concentration.	The variability of the automated method should be not more than 6.0% (RSD).
Dissolution	Six sample determinations at the nominal concentration.	The variability of the automated method should be not more than 6.0% (RSD) at the Q point.

should be considered at three levels: repeatability, intermediate precision, and reproducibility.

Repeatability. Repeatability expresses the precision under the same operating conditions over a short interval of time. The recommended testing for automated content uniformity, assays, degradation and impurity methods, and dissolution methods are listed in Table 5.1.

Intermediate Precision. Intermediate precision expresses within-laboratory variation and is generally performed on different days using different analysts, equipment, and sample preparations. This test may not be applicable if the laboratory has only one workstation. Additionally, this test may not be appropriate for automated workstations that are operating under the same environment and controls within a laboratory. This assumption is made on the basis that the automated workstations are identical (i.e., same configuration, same software and hardware) and that they have been suitably qualified and maintained to a consistent standard and operate under a similar climatic environment. The influence of the analyst is reduced to the preparation of solvents, and this should be covered by the robustness studies.

Reproducibility. Reproducibility expresses the precision between laboratories and would usually involve technical transfer of methods to laboratories in different geographical locations. The recommended testing for content uniformity, assays, degradation and impurity methods, and dissolution methods are listed in Table 5.2.

Table 5.2. Testing for Reproducibility

Method	Testing	Acceptance Criteria
Assay	Minimum of two automated runs with different analysts on different systems if possible. Twelve determinations total at the nominal concentration (six determinations from the same batch on each automated system).	The difference between the means of the individual runs is not more than 2.0%. The variability of the individual runs is not more than 2.0% (RSD) or less than the existing manual method, whichever is greater.
Degradation and impurity	Minimum of two automated runs with different analysts on different systems if possible. Four determinations total at the nominal concentration (two determinations from the same batch on each automated system). Batch should exhibit a typical impurity profile.	All impurities above LOQ are detected, and no additional peaks should be observed. Comparison of the impurities agree according to the following table:
Content uniformity	Minimum of two automated runs with different analysts on different systems if possible. Twenty determinations total at the nominal concentration (ten determinations from the same batch on each automated system). Batch should exhibit a typical impurity profile.	The difference between the means of the individual runs is no more than 2.0%. The variability of the individual runs is no more than 6.0% (RSD) or is less than the manual method, whichever is the greater.
Dissolution	Minimum of two automated runs with different analysts on different systems if possible. Twelve determinations total at the nominal concentration (one set of six samples from the same batch on each automated system). Batch should exhibit a typical impurity profile.	The difference between the means of the individual runs is no more than 6.0% at the Q point. The variability of the individual runs is no more than 6.0% (RSD) at the Q point or is less that of the manual method, whichever is greater.

Table within "Degradation and impurity" acceptance criteria:

Impurity Level	Absolute Agreement
$\leq 0.5\%$	$\pm 0.1\%$
$0.5-1.0\%$	$\pm 0.2\%$
$\geq 1.0\%$	$\pm 0.3\%$

Table 5.3. Testing for Accuracy

Method	Testing	Acceptance Criteria
Assay	Nine determinations total, three concentrations, three separately prepared replicates of each. Range of 80 to 120% of the nominal concentration; determine the recovery of active sample from the inert matrix.	All individual recoveries within 2.0% of the targeted concentrations.
Degradation and impurity	Two determinations total. Spike a placebo mix or solid dose form with available synthetic impurities and degradation products applicable to the method. Typically, spike at the specification limit.	All impurities above LOQ are detected and no additional peaks should be observed. Comparison of the impurities agree according to the following table:

Impurity Level	Absolute Agreement
$\leq 0.5\%$	$\pm 0.1\%$
$0.5-1.0\%$	$\pm 0.2\%$
$\geq 1.0\%$	$\pm 0.3\%$

Method	Testing	Acceptance Criteria
Content uniformity	Nine determinations total, three concentrations, three separately prepared replicates of each. Range of 70 to 130% of the nominal concentration, determine the recovery of active from the inert matrix.	All individual recoveries within 2.0% of the targeted concentrations.
Dissolution	Twelve determinations total, four concentrations, three separately prepared replicates of each. Range of 0%, 50%, 100%, and 120% of the nominal concentration; determine recovery of active sample from inert matrix.	All individual recoveries within 3.0% of the targeted concentrations.

Accuracy. The accuracy of the method expresses the closeness of agreement between the experimental result and the true value. When converting a validated manual method to an automated procedure, it may not be necessary to perform an accuracy study. Instead, it may be sufficient to rely on the comparison to manual data and on additional supporting validation on the automated procedure. When developing and validating the automated method as the first-intent method, it will be necessary to perform an accuracy study. The ICH guidelines describe several methods to determine the accuracy of a method [7]. For drug product assay methods, accuracy can be demonstrated by performing recovery experiments on synthetic mixtures of the product components to which known quantities of the drug have been added. If it is not possible to obtain all components of the drug product, another approach is the standard addition method, where the drug product is spiked with known quantities of active sample. A third approach is a comparison of the results from the automated method with the results from a second well-characterized procedure for which the accuracy has been established. Finally, although not recommended, accuracy can be inferred from precision, linearity, and specificity. The recommended testing for content uniformity, assays, degradation and impurity methods, and dissolution methods are listed in Table 5.3.

Equivalency. This test compares the results of the automated procedure with the results of the validated manual method. If accuracy of the automated procedure has been proven, it may not be necessary to perform the equivalency study. However, if the manual method does not exist, then accuracy and reproducibility data should be used to assess the suitability of the automated method. The recommended testing for content uniformity, assays, degradation and impurity methods and dissolution methods are listed in Table 5.4.

Robustness. Examples of typical possible sources of variation in automated methods are homogenization speed, homogenization time, age of sample, accuracy of solvent dispense, and temperature variation. If all studies described in the method development have been performed, the robustness of the sample preparation has been demonstrated and does not require additional testing. Parameters in relation to the measurement technique may need to be considered and are covered in the relevant chapter.

5.4 COMMON PROBLEMS AND SOLUTIONS

When attempting to convert a manual method into an automated method, there are certain elements, such as tablet size and solvent selection, which will have an impact on the ease of the conversion from manual to automated. For instance, some of the elements of an assay method that would make it easier to automate would be that the dosage form fits into a test tube; the extraction uses neutral media or acid not more concentrated than $0.1 \ M$; makes use of nonvolatile, low-toxicity solvents; does not use surfactants; and uses premixed, room-temperature solvents. Some of the elements of a dissolution method that would make it easier to automate would be that the dosage form fits in the sample carousel, does not use media more concentrated than $0.1 \ M$ acid, does not use isopropanol or surfactant in large quantities, uses magnetic sinkers or no sinkers at all, and uses no or minimal reagent addition volumes for pH control.

1. *Manual versus automated methods.* When difficulties arise in the development or validation of an automated method, it is often due to a minor difference

Table 5.4. Testing for Equivalency

Method	Testing	Acceptance Criteria
Assay	Six determinations at the nominal concentration, ideally from three different batches (18 automated determinations total).	The difference between the mean determinations of the manual and automated methods is 2.0% or less. The variability of the automated method is not more than 2.0% (RSD) or less than the existing manual method.
Degradation and impurity	Four determinations at the nominal concentration, ideally two determinations from two different batches.	All impurities above LOQ are detected, and no additional peaks should be observed. Comparison of the impurities agree according to the following table:

Impurity Level	Absolute Agreement
≤0.5%	±0.1%
0.5–1.0%	±0.2%
≥1.0%	±0.3%

Method	Testing	Acceptance Criteria
Content uniformity	Thirty determinations at the nominal concentration; ideally, 10 determinations from three different batches.	The difference between the mean determinations of the manual and automated methods is 2.0% or less. The variability of the automated method is not more than 6.0% (RSD) or less than the existing manual method.
Dissolution	Eighteen determinations at the nominal concentration; ideally, six determinations from three different batches.	The difference between the mean determinations of the manual and automated methods is 6.0% or less at the Q point. The variability of the automated method is not more than 6.0% (RSD) at the Q point.

between the manual and automated process that has not been compensated or adjusted for in the automated procedure. For example, during the automation of a dissolution method, the sequence of actions that are used to heat the media and fill the vessels in the automated method must be equivalent to the steps followed in the manual method. In a typical automated method, the media is heated to 37°C and then dispensed gravimetrically to the vessel, taking into account the density of the media at 37°C. However, in the manual method, the analyst will often measure the appropriate volume of media at room temperature, fill the vessels, and then heat in the vessel to 37°C. This difference in process may affect the dissolution results because the final volume in the vessels will differ.

In some cases, the automated method will yield far greater benefits than expected. Not only can an automated method yield large time savings, it can sometimes increase the reliability and precision of the results. For instance, when working with a product that has a high hydroxypropyl methylcellulose (HPMC) content, it is often difficult to dilute the volumetric flasks to the mark consistently because of the swelling and frothing caused by the HPMC in solution. Preparing the solutions gravimetrically avoids this issue, and thus the automated system yields samples that are prepared in a more reliable fashion, which in return produces results of higher consistency and precision.

2. *Solvent additions.* In the case of an assay/content uniformity/degradation and impurities method where the method is such that the sample is disintegrated in 50 mL of one solvent with subsequent addition of 20 mL of a different solvent prior to the extraction, two options present themselves for the automated method. The first option, which is preferred, is to investigate if full extraction can be achieved in the mixed solvent system with homogenization. Homogenization may assist in the dispersing of the tablets so that there may not be a need to add the solvents separately. In this case, the two solvents would be premixed, cooled to room temperature, and the density of the mixture would be determined. The automated process would simply add the required volume of the premixed solvents to the sample to proceed with the extraction. The second option, which is less desirable, is to follow the step indicated in the manual method and add the solvents separately. The volume change due to solvent mixing (solvent contraction or solvent expansion) is measured by weighing the volumetric flask with the first solvent and sample, then adding the second solvent and continuing the sample preparation until the sample is diluted to the mark. The sample flask is weighed again and the volume of the second solvent is calculated by taking the density into account. In this case, the volume displacement due to the sample is incorporated into the solvent mixing effect and does not need to be determined separately.

3. *Loss of active investigation.* When investigating system compatibility for active adsorption at the early stage of method development, an automated robust extraction method for the sample is not always available, and it is therefore difficult to assess loss of active in the system. It is best to verify for lost of active sample (adsorption) using a standard solution processed as a sample by the automated workstation. If recovery of the active is low, each section of the system with which the active comes in contact can be checked for adsorption separately. This can be achieved through manual sampling of the vessel before the solution comes in contact with the tubing, or by exposing sections of the tubing to the standard solution over a period of time and performing recovery experiments on the standard solution.

4. *Carryover.* If the automated method is coupled with online HPLC analysis and carryover is observed from one sample to the next, the HPLC injector must be considered as a potential source of carryover. Normally, if carryover is observed,

the vessel wash and the fluid path wash parameters would be modified to resolve the carryover and the solution with which the system is being washed may be changed. Since the HPLC injector may or may not be part of the fluid path, it is important to remember to isolate that particular section of the system and investigate it for carryover. This may constitute injecting a few blank samples (solvent) after each sample to verify that no active peak(s) are present in the blank sample chromatogram. Cumulative carryover should also be verified.

5. *Extraction study design.* When designing the homogenization process, it is recommended to perform multiple short pulses rather than a few long pulses, to avoid heating the solution. Soak/settle intervals between pulses can also help alleviate this problem. To confirm that the extraction process is complete, the homogenization vessel can be sampled after each extraction step and the results compared to see if further extraction was achieved after each extraction step. Alternatively, the extraction profile can be determined by creating various methods, each with an increasing number of extraction steps and analyzing the appropriate number of samples by each method.

6. *Solution stability.* When solution stability after a defined period of time fails due to evaporation of the solution, attempts to diminish the amount of evaporation can be made. Changing the cap on the test tube to a heavier and evaporation-resistant cap can sometimes reduce the speed of evaporation enough to allow for at least 24 h of solution stability. Using the largest volume possible in the test tube will minimize the effect of evaporation.

7. *Design of Validation Experiments.* For the validation of an automated assay method based on a validated manual method and involving a technical transfer of the method, the following experiments would be required:

- *Repeatability:* six determinations at nominal
- *Reproducibility:* six determinations from the same batch (batch A) on two systems (12 determinations total)
- *Equivalency:* six determinations from three different batches (batch A, B, C) (18 determinations total)

To maximize efficiency, the recommended approach is to perform the equivalency study first and use the data of batch A to fulfill some of the repeatability and reproducibility requirements. The remaining experimental work would be manual analysis of the samples for the equivalency study and the reproducibility study at the receiving laboratory.

The validation of an automated content uniformity method and an automated dissolution method based on a validated manual method can adopt a similar approach where the equivalency data can support the requirements of the other studies involved. When validating an automated method that is not based on a validated manual method, only the repeatability and reproducibility can be combined to reduce the amount of analysis to be done since the equivalency study will not be executed.

5.5 CONCLUSIONS

The key to successful integration of automation into the modern analytical laboratory is a sound approach to method development and validation. The end result of a well-developed and validated automated method will be a robust analytical method that should pay dividends by being able routinely to produce sound analytical data to support crucial regulatory submissions.

REFERENCES

1. R. J. Fix, Sr., J. M. Rowe, and B. C. McConnell, *Proceedings of the 1999 International Symposium on Laboratory Automation*, Zymark, Hopkinton, MA, p. 23, 1999.
2. M. Freeman et al., Positional paper on the qualification of analytical equipment, *Pharm. Technol. Eur.*, **7**(11), 40, 1995.
3. A. Walsh et al., Development and validation of automated methods for finished product testing, *Pharm. Technol. Eur.*, **24**(3), 134, 2000.
4. S. M. Han and A. Munro, Transfer from manual to automated sample preparation: a case study, *J. Pharm. Biomed. Anal.*, **20**, 785, 1999.
5. T. A. Steinman and E. Parente, Qualification of a Zymark® MultiDose® automated dissolution workstation for dissolution testing of fexofenadine HCl capsules, *Dissolut. Technol.*, May 2001.
6. V. Fuerte, M. Maldonado, and G. D. Rees, The multicomponent automated dissolution system: an alternative in the development and pharmaceutical analysis of generic polydrugs, *J. Pharm. Biomed. Anal.*, **21**, 267, 1999.
7. ICH Harmonized Tripartite Guideline, ICH Q2A; Text on Validation of Analytical Procedures, June 1997.
8. FDA Guidance for Industry, *Dissolution Testing of Immediate Release Solid Dosage Forms*, U.S. Department of Health and Human Services, Food and Drug Administration, Center for Drug Evaluation and Research, Washington, DC, pp. 8–10, Aug. 1997.
9. United States Pharmacopeia, General Chapters: Physical Tests and Determinations: <711> Dissolution, Official Compendia of Standards, USP 25, NF 20, p. 2011, 2002.

6

ANALYSIS OF PHARMACEUTICAL INACTIVE INGREDIENTS

Xue-Ming Zhang, Ph.D.
Novex Pharma

6.1 INTRODUCTION

Pharmaceutical dosage forms contain active and inactive ingredients usually called *excipients*. The effectiveness of the dosage form depends on the manufacturing process and the interrelationship between excipients and their impact on the active ingredients. There are many dosage forms, such as liquids, semisolids, and solids. The excipients used in each dosage form can be varied from simple inorganic salts to complex organic polymers. This chapter focuses on (1) the commonly used excipients in liquid formulations, (2) the selection of analytical methods to quantify excipients, (3) the suggested validation elements, and (4) common problems and solutions. The concept that is used for the analysis and validation of excipients in the simple liquid formulation can be applied to any excipients that are used in other types of formulations. Excipients that utilize compendial standards will also apply the compendial methodologies after the appropriate verification.

6.2 COMMONLY USED EXCIPIENTS IN LIQUID MEDICINES

Liquid medicines generally include oral liquids, suspensions, emulsions, inhalations, nasal solutions and suspensions, topical semisolids and topical liquids, ophthalmics, and parenterals. There are numerous excipients used for liquid

Analytical Method Validation and Instrument Performance Verification, Edited by Chung Chow Chan, Herman Lam, Y. C. Lee, and Xue-Ming Zhang
ISBN 0-471-25953-5 Copyright © 2004 John Wiley & Sons, Inc.

Table 6.1. Commonly Used Excipients in Liquid Medicines

Function	Typical Excipients
Antimicrobial preservative	Ethanol, benzoic acid, sorbic acid, the hydroxybenzoate esters, phenylethyl alcohol, and glycerin
Antioxidant	Ascorbic acid, citric acid, sodium metabisulfite, and sodium sulfite
Suspending agent	Carbomer, carmellose, microcrystalline cellulose, sodium carboxymethylcellulose, povidone, sodium alginate, tragacanth, and xanthan gum
Emulsifying agent	Acacia and methylcellulose, glycerol esters, polysorbates and sorbitan esters, fatty acids, sodium stearate, carbomer Macrogol esters, polyvinyl alcohol, and glycerides
Flavoring agent	Juices extracts (liquorize), spirits (orange, lemon), syrups (black currant), tinctures (ginger), and aromatic waters
Coloring agent	Mineral pigments (iron oxides), natural colorants, anthocyanins, carotenoids, chlorophylls riboflavine, red beetroot extract, and caramel; synthetic organic dyes azo compounds
Sweetening agent	Sugars, including glucose, sucrose syrup, and honey, sorbitol, mannitol, and xylitol, sodium and calcium salts of saccharin, aspartame, potassium thaumatin
Viscosity-enhancing agent	Sugars and polyvinyl alcohol, povidone, and cellulose
Flocculating agent	Electrolytes, polymers, starch, sodium alginate, and carbomer
Wetting agent	Polysorbates or sorbitan esters, acacia, and tragacanth
Gelling agent	Aluminum magnesium silicate, bentonite, carbomers, cellulose derivatives, gelatin, pectin, polyvinyl alcohol, alginates, starch, and xanthan gum
Buffering agents	Borates, citrates, acetates, and phosphates

Source: Refs. 1 and 2.

formulations. Table 6.1 summarizes the types and functions of most commonly used excipients.

6.3 METHOD SELECTION

Prior to selecting an analytical method, the excipients must be identified. Several sources of references can be used to obtain the required information [3,4]. Method development and validation strategies are based primarily on the type and amount of excipients in the formulation. Generally speaking, there are two categories of analytical methodologies available to the analyst: specific and non-specific. A specific analytical method is designed to separate potential interferences and to test only for the analyte of interest. Chromatographic methods such as high-performance liquid chromatography (HPLC), gas chromatography (GC), HPLC-MS (mass spectrography), GC-MS, and sometimes ion chromatography (IC) are examples of specific analytical method. A nonspecific method measures a property of the sample solution rather than testing discretely for the analyte in the solution. Conductivity, water content, pH, and gravimetric analysis are examples of nonspecific test methods. Although UV-Vis and atomic absorption methods have some element of specificity, they are more susceptible to interferences than chromatographic method.

In this chapter, analysis of the excipient in Nasonex (mometasone naral spray) is used as a typical example to demonstrate how to select a proper analytical method. Nasonex Nasal Spray contains Mometasone Furoate, microcrystalline cellulose, and carboxymethylcellulose sodium NF, citric acid USP, sodium citrate USP, benzalkonium chloride solution NF, glycerin USP, polysorbate 80 NF, phenylethyl alcohol USP, and water USP.

6.3.1 Quantitation of Citric Acid

Citric acid is one of the low-molecular-weight (LMW) organic acids. In Nasonex it is used as a buffering agent. The usual methods for LMW organic acid analysis include capillary gas chromatography after solvent extraction and derivatization. The derivatizing techniques used are (1) formation of methyl esters using BF3−methanol, and (2) formation of butyl ester using HCl−butanol. Another routine method for the analysis of LMW organic acid is HPLC. Organic acids have been analyzed using both normal- and reversed−phase separations. Ion chromatography with conductivity or electrochemical detector has also been adopted for LMW organic acid analysis. The total citrate (citric acid and sodium citrate) in Nasonex is quantified by a reversed-phase HPLC method. The HPLC system is equipped with a UV−Vis detector, a Supelcosil LC-18DB column, 250 mm × 4.6 mm, 5 μm. The mobile phase consisted of 10% methanol and 90% buffer containing 0.01 M tetramethylammonium hydroxide with pH 2.5. Sample was diluted with 1 N HCl. Equal volumes (100 μL) of standard and sample solutions were injected to the HP1100 system. The peak of citric acid was monitored by a spectrophotometric detector at 210 nm. The total amount of citrate was found to be 3.66 g/kg. The pH value of Nasonex was accurately measured by a pH meter (pH 4.76) The concentration of citric acid and sodium citrate were then calculated based on the Henderson−Hasselbalch equation:

$$\text{pH} = \text{p}K_a + \log \frac{C_{\text{salt}}}{C_{\text{acid}}}$$

where C_{salt} and C_{acid} are the molar concentrations of the salt form and acid form, respectively. The following example illustrates the calculation of the concentrations of citric acid and sodium citrate.

$$\text{pH of Nasonex} = 4.76$$

$$\text{p}K_{a_2} \text{ of citric acid} = 4.76$$

The molar concentration ratio required to balance the Henderson–Hasselbalch equation:

$$4.76 - 4.76 = 0$$

Since $\log 1 = 0$, the molar concentration ratio of citric acid and sodium citrate is $1:1$. In 3.66 g/kg of total citrate citric acid and sodium citrates, each accounts for 50%. That is, 1.83 g of citric acid and 2.8 g of sodium citrate (1.83 g × 294.1/192.1 = 2.8 g).

6.3.2 Quantitation of Glycerin

Glycerin is used in Nasonex primarily as a humectant. For its quantification, both capillary gas chromatography method and HPLC methods may be selected. The GC is equipped with a flame-ionization defector, a 0.53 mm × 30 m fused silica analytical column coated with 3.0-μm G43 stationary phase, and a 0.53 mm × 5 m silica guard column deactivated with phenylmethyl siloxane. The carrier gas was helium with a linear velocity of about 35 cm/s. The injection port and detector temperature was maintained at 240 and 260°C, respectively. The injection mode is splitless. The column temperature is programmed to be maintained at about 40°C for 20 min, then to increase to 250°C at a rate of 10°C/min and to hold at 250°C for 15 min.

A standard solution was prepared in 50% acetonitrile and water having a known concentration of about 100 μg of glycerin per milliliter. Then 1.0 g of Nasonex was transferred into a 200-mL volumetric flask. About 30 mg of calcium chloride was added and diluted to volume with 50% acetonitrile in water and mixed. A portion of standard and sample solutions were filtered through a 0.45-μm nylon syringe filter. Equal volumes (1 μL) of standard and sample solutions were injected separately into the chromatographic system.

The HPLC was equipped with a refractive index detector, and a 150 mm × 4.6 mm, 3-μm Phenomenex amino column. The mobile phase was 70% acetonitrile in water. The flow rate was 1 mL/min. Equal volumes (25 μL) of standard and sample solutions were injected separately into the system. The amount quantified by capillary GC and HPLC methods were fairly comparable (about 21 g/kg) and both methods were specific, precise, and accurate. However, the total run time for HPLC method was less than one-third that of the GC method.

6.3.3 Quantitation of Benzalkonium Chloride

Benzalkonium chloride (BAC) is used as an antimicrobial preservative in Nasonex. It is quantifiable using a HPLC method with a column that has a cyano group chemically bonded to porous silica particles. The HPLC was equipped with a UV–Vis detector, a 150 mm × 4.6 mm, 3-μm Spherisorb S_3 CN column. The mobile phase consisted of 45% acetonitrile in 0.05 M phosphate buffer (pH 6.0). The flow rate was set at 1 mL/min. The following procedure was used for sample and standard preparations. About 2.5 g of Nasonex was transferred into a 25-mL volumetric flask. This was diluted to volume with acetonitrile and mix. The standard solution was prepared at approximately 0.025 mg/mL. Equal volumes (25 μL) of standard and sample solutions were injected into the system. The detector wavelength was set up at 210 nm. Peak response was compared with that of standard and the amount (g/kg) of benzalkonium chloride in sample solution was calculated.

6.3.4   Quantitation of Polysorbate 80

Polysorbate 80 is widely used as a nonionic surfactant in liquid pharmaceutical products such as inhalation, suspension, and nasal suspension products, due to its properties of solubilization, reduction of surface and interfacial tension, and wetting. Direct analysis of Polysorbate 80 is quite time consuming. Size-exclusion chromatography (SEC) has been reported [5] in which a mobile phase contained the surfactant at concentrations above the critical micelle concentration. Polysorbate 80 appeared as a very broad peak and coeluted with other peaks, which makes quantification in Nasonex impossible.

Fatty acids can be determined easily by GC and HPLC. Hydrolysis of polysorbate 80 will release oleic acid. Thus, if a suitable hydrolysis procedure can be established to release oleic acid quantitatively, either a GC or an HPLC technique can be performed to determine the amount of polysorbate 80 in Nasonex. Approximately 9 g of Nasonex and an equivalent amount of polysorbate 80 standard were transferred into two separate flasks, followed by the addition of 1.0 mL of 1 N NaOH, and mixed well. The mixture was heated in a water bath at 70°C for 3 h. The sample and standard solutions were removed from the water bath and cooled to room temperature. An equal volume of acetonitrile was added to the solutions, mixed well, and centrifuged. The supernatant portion is filtered through a 0.2-μm nylon syringe filter. The hydrolysis can also be carried out under acidic conditions. Most of the procedures in the acidic conditions were similar to basic hydrolysis. The only difference was adjustment of the final H_2SO_4 concentration to 1 N and extending the hydrolysis time to 24 h.

The HPLC was equipped with a UV–Vis detector (VWD = 210 nm), a water symmetry C_{18} column, 150 mm × 3.9 mm, 5 μm. The mobile phase was 25% acetonitrile in 75% 20 mM phosphate buffer (pH 2.8). The flow rate was set at 1.5 mL/min. Equal volumes (25 μL) of standard and sample solutions were injected into the chromatographic system. Quantification of polysorbate 80 is based on a comparison of the response of oleic acid in sample and that of oleic acid in standard solution [6]. When polysorbate 80 was quantified by GC, the released oleic acid can be detected without derivatization and prepared according to the HPLC method.

The chromatographic procedure was carried out using a 0.53 mm × 30 m fused silica analytical column coated with 1.0 μm Supelcowax 10 stationary phase, a flame-ionization detector, and a split injector (1 : 30). The carrier gas was helium at a flow rate of 3 mL/min. The temperature of the injection port and detector was set up at 250 and 280°C, respectively. The column temperature program was programmed as follows:

Time (min)	Temperature (°C)	Rate (°C/min)
0–1	200	–
1–20	280	4
20	280	–

Equal volumes (1 μL) of standard and sample solutions were injected into the system and the amount present in the sample solution was calculated according to the HPLC method. Equivalent results were obtained from the HPLC and GC methods. If a product contains a lower concentration of Polysorbate 80 (e.g., lower than 0.05 g/kg), the GC method may be superior over the HPLC method. In such a case, derivatization and extraction steps may be required during sample preparation. The yield of oleic acid from hydrolysis of Polysorbate 80 was examined by analyzing a Polysorbate 80 spiked sample against an oleic acid standard. The calculation was based on the fact that 1 mol of polysorbate 80 releases 1 mol of oleic acid.

6.3.5 Quantitation of Microcrystalline Cellulose and Carboxymethylcellulose

The total cellulose (microcrystalline cellulose and carboxymethylcellulose sodium) quantitation is achievable through use of one of the following: (1) total solid subtraction, (2) indirect quantification of sodium carboxymethylcellulose (NaCMC), or (3) direct quantification of the hydrolyzed monosaccharide.

There are three ways to obtain the total solid: (1) determine the water content by Karl Fisher titration, (2) weigh an evaporate portion of product, or (3) sum up all the ingredients, including mometasone furoate and phenylethyl alcohol, which are stated (claimed) on the label. The combination of all three measurements generally gives quite an accurate estimation of the total solid and total cellulose:

total cellulose (g/kg) = total solid (g/kg)

– sum of all other ingredients in Nasonex except the cellulose (g/kg)

Avicel RC and CL are water-dispersible, colloidal, microcrystalline cellulose products made for use in liquid preparations. Avicel RC and CL are coprocessed mixtures of microcrystalline cellulose and sodium carboxymethylcellulose. The amount of NaCMC can be determined using the IC method. About 10 g of Nasonex and about 25 mg of NaCMC NF are separately refluxed with 30 mL of glacial acetic acid for 2 h. The refluxed mixture is transferred to a 100-mL volumetric flask and diluted to volume with purified water, and mixed. The ion chromatograph (IC) was equipped with a suppressed conductivity detector, a 4-mm CSRS suppressor, current at 50 mA, a 250 mm × 4-mm Ion Pac CS 12A column and a 50 mm × 4 mm Ion Pac CG 12A guard column. The mobile phase is 0.13% methanesulfonic acid in water with a flow rate of 1 mL/min. Equal volumes of NaCMC standard and sample solutions as well as a sodium standard solution are injected into the IC system.

The total sodium in a sample solution is calculated based on the sodium standard. The sodium contributed by NaCMC in a sample solution is the total sodium minus the sodium from sodium citrate (corrected sodium). The amount of NaCMC in a sample can be calculated against standard NaCMC using the corrected sodium. The calculation of total cellulose in the sample solution is then as follows:

$$\text{Avicel RC-591 (g/kg)} = \frac{\text{NaCMC} \times 100}{11.05}$$

$$\text{Avicel CL-611 (g/kg)} = \frac{\text{NaCMC} \times 100}{15.05}$$

where 11.05 and 15.05 are the average concentrations of NaCMC specified in the Avicel RC-591 and Avicel CL-611 specifications, respectively. Nasonex may contain about 20 g/kg of Avicel RC-591 or about 14.7 g/kg of Avicel CL-611. However, if the total solid has been taken into consideration, it was confirmed that 20 g/kg of Avicel RC-591 is the only correct choice.

6.4 VALIDATION PRACTICE

Most of the analytical methods developed in the laboratory for quantitation of excipient are specific methods, usually HPLC, GC, and IC methods. The raw material of an individual excipient is used directly as the standard to quantify the excipient in the formulation. The concentration of the standard varies with the mode of detection and formulation. It ranges from low micrograms to high milligrams per milliliter. The validation elements required depend on the type and origin of the method. It is acceptable to design the experimental work so that the appropriate validation characteristics can be considered simultaneously to provide sound overall knowledge of the capabilities of the test methods. Unlike the analysis of the active pharmaceutical ingredient (API), neither the FDA or nor the ICH [7–9] has clear requirements for the analysis of excipients.

6.4.1 Specificity

The specificity tests depend on the type and purpose of the method. If a specific method is being validated, an interference study should be undertaken. In the case above of the analysis of citric acid and sodium citrate, all other ingredients except microcrystalline cellulose and carboxymethyl cellulose sodium should be chromatographed separately. The known impurities related to memetasone furoate, if available, should be injected as well as the diluting solvent. Interference from filters is required. For nonspecific methods, specificity studies must be determined on a case-by-case basis. For example, the determination of molar ratio of citric acid and sodium citrate by pH in Nasonex should be designed to exclude any other contribution of acidity from other ingredients.

Table 6.2. Method Accuracy (Percent Recovery) for Polysorbate 80

50% Spiked Samples (0.09845 mg/mL)	100% Spiked Samples (0.19690 mg/mL)	150% Spiked Samples (0.29535 mg/mL)
98.0	100.1	100.8
99.5	101.2	100.7
100.6	100.7	101.9
Average 99.4	Average 100.7	Average 101.1
1.3% RSD	0.6% RSD	0.7% RSD

Source: Ref. 6.

6.4.2 System Precision

Inject the excipient standard solution at the working concentration specified in the method. The relative standard deviation (% RSD) of five or six consecutive injections is NMT 2%. The requirement for % RSD may be modified depending on the properties of the analyte and the test method. In the analysis of Polysorbate 80 in Nasonex, 5% RSD for system precision was acceptable, due to its very low concentration in the formulation.

6.4.3 Method Precision

Perform the assay of six individual samples on a single day according to the method proposed with NMT 2% RSD. In most cases, a duplicate assay of the second lot should also be made to check the lot-to-lot variation.

6.4.4 Accuracy

The accuracy of the method related to the excipient is achieved by assay of three preparations at 50%, 100%, and 150% of the proposed concentration by spiking the excipient. The mean recovery at each level should be between 97 and 103% (Table 6.2).

6.4.5 Linearity

A minimum of 5 points in the analyte range 50 to 150% is acceptable. A correlation coefficient of 0.995 or better is a reasonable expectation (Figure 6.1). The % RSD of the average response factor should be NMT 5%.

6.5 COMMON PROBLEMS AND SOLUTIONS

The most common problem encountered during method validation is low recovery, especially for very low concentrations. There are two potential causes

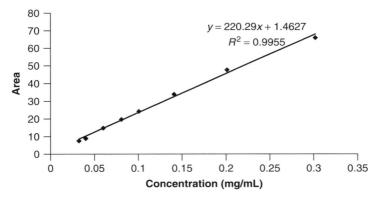

Figure 6.1. Linearity of polysorbate 80.

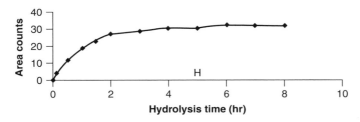

Figure 6.2. Hydrolysis of polysorbate 80 with 0.1 N KOH.

of low recovery: (1) incorrect sample solvent was used, which gave incomplete extraction of the substrate, or (2) the sample preparation procedure was not appropriate, causing loss of the analyte. The following points should be considered when investigating the cause of low recovery: (1) increasing the solvent strength of the sample solvent or changing the pH value of the extraction solvent, (2) increasing the shaking time or more vigorous shaking to assist in extracting the analyte from the substrate, or (3) the use of sonication to increase extraction of the analyte from the substrate. Special attention should be given to analytes that have a high vapor pressure or low melting point or that are thermo- or light-sensitive. Due to the low concentration and thermosensitivity of polysorbate 80, special attention was paid to (1) the amount of sample used, (2) the final concentration of the sodium hydroxide, (3) the temperature and time of hydrolysis (Figure 6.2), and (4) the strength of the extraction solvent.

REFERENCES

1. H. Kibbe, ed., *Handbook of Pharmaceutical Excipients*, 3rd ed., Pharmaceutical Press, London.

2. The Pharmaceutical CODEX, *Principles and Practice of Pharmaceutics*, 12th ed., Pharmaceutical Press, London, 1994.

3. *Physicians' Desk Reference* 54th ed. Medical Economics Company, Montvale, NJ, p. 2819, 2000.

4. *Compendium of Pharmaceuticals and Specialties*, Canadian Pharmacists Association, 2003.

5. T. H. Tani, J. M. Moore, and T. W. Patapoff, *J. Chromatogr. A*, **786**, 99–106, 1997.

6. M. Hu, M. Neculescu, X. M. Zhang, and A. Hui, *J. Chromatogr. A*, **984**(2), 233–236, 2003.

7. Guidance for Industry, *Analytical Procedures and Methods Validation Chemistry, Manufacturing, and Controls Documentation*, Aug. 2000.

8. ICH Harmonized Tripartite Guideline, ICH Q2A, *Text on Validation of Analytical Procedures*, Mar. 1995.

9. ICH Harmonized Tripartite Guideline, ICH Q2B, *Validation of Analytical Procedures: Methodology*, May 1997.

7

VALIDATION STUDY OF JP HEAVY METAL LIMIT TEST

Yoshiki Nishiyama
Eli Lilly Japan K.K.

7.1 INTRODUCTION

In this chapter we describe characteristic validation procedures of the Heavy Metals Limit Test in the *Japanese Pharmacopoeia* (JP) [1]. Although an equivalent test is commonly listed in both the *United States Pharmacopoeia* and the *European Pharmacopoeia*, there are differences in the color reagents and conditions of sample preparation of the JP procedure. Heavy metals are defined in the JP as poisonous metallic impurities such as Pb, Bi, Cu, Cd, Sn, and Hg that form colored colloidal precipitates with sodium sulfide TS in a slightly acidic solution of pH 3 to 4. The level is expressed as the equivalent quantity of lead.

In JP monographs, the specification value and testing procedure are described as a comparative limit test for the quantity of heavy metals that exist in drug substances as inorganic impurities. The permissible limit on the "ppm" scale for heavy metals (as Pb) is prescribed as the specification value. From the development stage to the establishment of the limit test method for a monograph, the validation in place has to be done for both the quantitative method for actual measurement and the comparative limit method with the control solution. However, the quantitative procedure has not been stated in the section on the heavy metals limit test in the JP's General Tests and other Japanese official

Analytical Method Validation and Instrument Performance Verification, Edited by Chung Chow Chan, Herman Lam, Y. C. Lee, and Xue-Ming Zhang
ISBN 0-471-25953-5 Copyright © 2004 John Wiley & Sons, Inc.

guidelines, even though the validation package for the actual value measurement is usually required as the rationale of limit value described in regulatory submission documents. Therefore, in this chapter we describe the approaches of validation studies that are required to establish and implement practices concerning the quantitation and the limit testing methods for heavy metals.

We focus on the requirements for registration in Japan. Thus, we provide information that considers the peculiarities of the requirements for the JP heavy metals test at the method development process. The compendial descriptions and relevant ICH guidelines in relation to this chapter are listed in the reference section. The principles for the validation requirements discussed in this chapter can be applied to heavy metal testing in general.

7.2 SCOPE OF CHAPTER

In the process for validation studies for developing the limit test of heavy metals according to JP requirements for the specifications of new drug substances, there are usually two significant individual procedures: (1) validation for the quantitation of heavy metals, and (2) validation for the limit test of heavy metals. Validation of the quantitation for heavy metals test assesses the applicability of a quantitation method where the measured values serve as the basis for the limit test used for the specification of heavy metals in the drug substance. This method determines the concentration of lead by a UV–Vis spectrophotometer using a calibration curve. These data should be described in regulatory submission documents as a piece of essential information for a new drug application. We follow with a discussion on characteristics and procedures of validation for developing the limit test specified in "Specifications and Testing Methods" of the drug substance.

7.3 VALIDATION PRACTICES

In this chapter, validation studies for developing the quantitative and JP limit test methods for heavy metals described in "Specifications and Testing Methods" of the drug substance section are described. Procedures to evaluate validation characteristics concerning the validation items of each test method are described. Table 7.1 summarizes the validation characteristics required by JP for testing of impurities.

7.3.1 Validation Procedures of Quantitation of Heavy Metals

The quantitation of heavy metals in a drug substance is a method to determine the quantity of lead by determining the intensity of coloring with sodium sulfide based on the absorbance at 400 nm by spectrophotometry. It is a prerequisite to set up the limit test according to the JP requirements. With regard to validation characteristics of the quantitation method, specificity is not required because of

Table 7.1. JP Validation Characteristics Requirements[a]

Characteristic	Identification	Testing for Impurities		Assay (dissolution; content/potency)
		Quantitation	Limit	
Accuracy	−	+	−	+
Precision				
Repeatability	−	+[b]	−	+[b]
Intermediate precision	−	−	−	+
Specificity[c]	+	+	+	+
Detection limit	−	−[d]	+	−
Quantitation limit	−	+	−	−
Linearity	−	+	−	+
Range	−	+	−	+

[a]−Signifies that this characteristic is not normally evaluated; + signifies that this characteristics is normally evaluated.
[b]Where reproducibility has been established, intermediate precision is not needed.
[c]Lack of specificity of one analytical procedure could be compensated by supporting analytical procedure(s).
[d]May be needed in some cases.

the definition of heavy metals as described in "Heavy Metals Limit Test" section of JP general testing method 22. Consequently, the following four validation characteristics should be evaluated: (1) accuracy, (2) precision (repeatability), (3) range and linearity, and (4) quantitation limit. Accuracy and precision are evaluations on the appropriateness of procedures adopted from the test solution preparation of methods 1 to 4 in JP. The sample content must be evaluated so that it is within the linearity, range, and quantitation limit.

Accuracy. In the quantitative method that is used to measure the heavy metal quantity in the drug substance, the accuracy is usually represented by the recovery rate obtained from a spiked recovery test where lead is added to the samples. Since the heavy metals limit test specified in monograph specifications is a test where the intensity of coloring of the samples with sodium sulfide is compared with that of the control solution, it is necessary to confirm that heavy metal components can be detected fully in the process of test solution preparation. The "Heavy Metals Limit Test" in JP specifies four preparation methods for the test solutions. An appropriate method will be selected and used for further testing. The test method that gives the best recovery rate is to be adopted. The procedure is as follows:

Procedure
    1. Add a volume of standard lead solution containing each of the following three levels of lead to samples of the drug substance: the content equal to the expected specification limit and those higher and lower than the

median level. At least three spiked samples should be prepared for each lead content.

2. Treat the three-level lead-spiked drug substance samples according to methods 1 to 4 to prepare the test solutions and the control solution. Separately, designate a solution prepared according to the same preparation method as the control solution. Using this control solution, determine the absorbance of the test solutions and calculate the recovery rate at each amount added.

A typical example of accuracy (recovery rate) is shown in Table 7.2. The samples were prepared by adding a standard lead solution corresponding to 2, 5, 10, and 15 ppm to 2.0 g of a compound. Results of the recovery were determined from the absorbance at 400 nm.

Precision (Repeatability). To evaluate the repeatability as specified in the quantitative method of heavy metals in the drug substance, the drug substance samples are treated according to the test solutions and the control solutions preparation method selected from methods 1 to 4 of "Heavy Metals Limit Test" in JP. Take five or six aliquot samples collected from a single lot of homogeneous drug substance and determine the quantity of heavy metal in each sample aliquot using the prepared test and control solutions. The data obtained are statistically analyzed.

Procedure
1. Take five or six aliquot samples from a single lot of homogeneous drug substance. The sample amount is equal to that specified in the limit test method in the "Specifications and Testing Methods."
2. Prepare the test and control solutions from the aforementioned sample according to the method of test solutions selected and the control solutions preparation procedure specified in the limit test method in the "Specifications and Testing Methods." Use these solutions to determine the quantity of heavy metal by the same calibration curve from the accuracy experiment. This procedure is repeated for each test solution.
3. Perform statistical analysis to obtain the standard deviation and relative standard deviation.

Table 7.2. Recovery of Heavy Metals (%)

Replicate	Amount of Lead Added (ppm)			
	2	5	10	15
1	85.2	98.5	93.8	93.6
2	98.9	91.6	91.7	97.2
3	98.9	93.0	88.3	85.8
Average	94.3	94.4	91.2	92.2

Table 7.3. Testing for Repeatability

Sample	Value Observed (ppm)	Mean	SD	RSD (%)
1	1.6			
2	1.7			
3	1.9	1.8	0.11	6.1
4	1.7			
5	1.9			
6	1.7			

A typical example of precision (repeatability) is shown in Table 7.3. The samples were prepared from 2.0 g of a compound. The data obtained were determined from the absorbance at 400 nm.

Range and Linearity. The range for which the linearity of the absorbance versus lead concentration calibration curve applied for the heavy metal quantitative method has to be confirmed. A wide range from ca. 1 ppm to two times the specification limit value (e.g., 15 or 20 ppm concentration relative to the sample weight) should be used for the evaluation range of the calibration curve. In general, it is recommended to use at least five concentration levels at regular intervals by using lead standard solution within the range studied. Actual data obtained are analyzed statistically by obtaining a regression line using the least squares method to evaluate linearity within the range specified. The linearity of the calibration curve is evaluated by the correlation coefficient, *y*-intercept, and slope of the regression line. A plot of the regression line should be included.

Procedure
1. Place an appropriate volume of standard lead solution which corresponds to each of five concentration levels in a Nessler tube, and add water to make up to 40 mL. Add 2 mL of dilute acetic acid and water to make up to 50 mL, and designate it as the test solution. Three tubes of test solution at each concentration level are prepared for separate runs.
2. The control solution is prepared by pipetting 2 mL of dilute acetic acid in a Nessler tube of the same material with the test solution and adding water to make up to 50 mL.
3. Add a drop of sodium sulfide to the test and control solutions, mix, and leave for 5 min. Then determine the absorbance of the test solution using the control solution as the control value.
4. Determine the absorbance at each concentration, and obtain the regression curve of lead concentration versus absorbance by the least squares method.

Figure 7.1 illustrates the typical linearity obtained for this test. The samples were prepared by adding standard lead solution corresponding to 2, 5, 10, and 15 ppm to 2.0 g of a compound. The response was linear within the range.

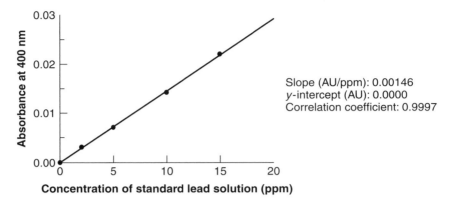

Figure 7.1. Linearity of calibration curve.

Quantitation Limit. When the quantity of heavy metals is determined from the calibration curve, it is recommended to estimate the lowest value of heavy metals concentration as the quantitation limit. The methods to estimate the quantitation limit are described in JP and ICH Q2B Guidelines [2], and an appropriate method should be selected from among these methods. An estimation of the quantitation limit can be obtained from the standard deviation of measured values of the low-concentration test solution. The standard deviation of background noise will be used to estimate a value for the standard signal-to-noise ratio (10 : 1).

Procedure
  1. Prepare five to six samples of the test solution with known low-lead concentration that is close to the quantitation limit, and determine the absorbance at 400 nm by spectrophotometer.
  2. Calculate the standard deviation of the samples above. Use this standard deviation as the standard deviation of background noise.
  3. Assuming that the standard deviations of the measured heavy metals values maintain the linearity up to the limit value, the quantitation limit (QL) corresponding to the signal-to-noise ratio of 10 : 1 (RSD = 10%) is calculated using the following formula:

$$QL = SD \times 10$$

where SD is the standard deviation of the absorbance of the test solution with known low-lead concentration that is close to the quantitation limit.

Using this procedure, the estimated quantitation limit is obtained from the variation of actual measurement values of low-concentration test solutions instead of background noise. However, the actual value of quantitation limit needs to be verified by experiment.

7.3.2   Validation Procedures of Limit Test of Heavy Metals

The validation of heavy metals limit test method is required to evaluate the specificity and the detection limit of the limit test of impurities. However, since the JP method is capable of detecting Pb, Bi, Cu, Cd, Sn, and Hg, specificity evaluation for these metals is not required. The detection limit of the method should be evaluated visually.

Detection Limit.   The detection limit of heavy metals limit test method is obtained from the test solutions and the control solution. These solutions are prepared using one of methods 1 to 4 of the control solution preparation method. The detection limit is determined by visual inspection of a series of diluted standard lead solutions.

Procedure
  1. Adjust the volume of lead standard solution to be the same as that for preparation of the control solution. Prepare at least seven or eight levels of lower-concentration test solutions from the specification limit value of 0.1 to 0.2 ppm (single replicate).
  2. Add a drop of sodium sulfide TS to the test solutions, mix, and hold for 5 min. Put the Nessler tube containing the test solution of each lead concentration on a white background. Visually observe it from the upper side. Determine the lowest concentration where darkening can be visually observable and designate it as the detection limit.

7.3.3   Preparation of Test Solutions and Control Solutions

The following are citations from the section on the preparation of test solutions and control solutions in JP 14, "Heavy Metals Limit Test." Unless otherwise specified, test solutions and control solutions are prepared as directed below:

Method 1.   Place an amount of the sample, directed in the monograph, in a Nessler tube. Dissolve in water to make 40 mL. Add 2 mL of dilute acetic acid and water to make 50 mL, and designate it as the test solution. The control solution is prepared by placing the volume of standard lead solution directed in the monograph in a Nessler tube, and adding 2 mL of dilute acetic acid and water to make 50 mL.

Method 2.   Place an amount of sample, directed in the monograph, in a quartz or porcelain crucible, cover loosely with a lid, and carbonize by gentle ignition. After cooling, add 2 mL of nitric acid and 5 drops of sulfuric acid, heat cautiously until white fumes are no longer evolved, and incinerate by ignition between 500 and 600°C. Cool, add 2 mL of hydrochloric acid, evaporate to dryness on a water bath, moisten the residue with 3 drops of hydrochloric acid, add 10 mL of hot water, and warm for 2 min. Then add 1 drop of phenolphthalein TS, add ammonia TS dropwise until the solution develops a pale red color, add 2 mL of dilute acetic acid, filter if necessary, and wash with 10 mL of water. Transfer the filtrate and washings to a Nessler tube, and add water to make 50 mL. Designate it as the test solution.

The control solution is prepared as follows: Evaporate a mixture of 2 mL of nitric acid, 5 drops of sulfuric acid, and 2 mL of hydrochloric acid on a water bath, further evaporate to dryness on a sand bath, and moisten the residue with 3 drops of hydrochloric acid. Proceed as directed in the test solution, then add the volume of standard lead solution directed in the monograph and water to make 50 mL.

Method 3. Place an amount of the sample, directed in the monograph, in a quartz or porcelain crucible, heat continuously, gently at first, and then increase the heat until incineration is completed. After cooling, add 1 mL of aqua regia, evaporate to dryness on a water bath, moisten the residue with 3 drops of hydrochloric acid, add 10 mL of hot water, and warm for 2 minutes. Add 1 drop of phenolphthalein TS, add ammonia TS dropwise until the solution develops a pale red color, add 2 mL of dilute acetic acid, filter if necessary, wash with 10 mL of water, transfer the filtrate and washing to a Nessler tube, and add water to make 50 mL. Designate it as the test solution.

The control solution is prepared as follows: Evaporate 1 mL of aqua regia to dryness on a water bath. Proceed as directed for the test solution, and add the volume of standard lead solution directed in the monograph and water to make 50 mL.

Method 4. Place an amount of the sample, directed in the monograph, in a platinum or porcelain crucible, mix with 10 mL of a solution of magnesium nitrate hexahydrate in ethanol (95) (1 in 10), fire the ethanol to burn, and carbonize by gradual heating. Cool, add 1 mL of sulfuric acid, heat carefully, and incinerate by ignition between 500 and 600°C. If a carbonized substance remains, moisten with a small amount of sulfuric acid, and incinerate by ignition. Cool, dissolve the residue in 3 mL of hydrochloric acid, evaporate on a water bath to dryness, wet the residue with 3 drops of hydrochloric acid, add 10 mL of water, and dissolve by warming. Add 1 drop of phenolphthalein TS, add ammonia TS dropwise until a pale red color develops, then add 2 mL of dilute acetic acid, filter if necessary, wash with 10 mL of water, transfer the filtrate and the washing to a Nessler tube, add water to make 50 mL, and use this solution as the test solution.

The control solution is prepared as follows: Take 10 mL of a solution of magnesium nitrate hexahydrate in ethanol (95) (1 in 10), and fire the ethanol to burn. Cool, add 1 mL of sulfuric acid, heat carefully, and ignite between 500 and 600°C. Cool and add 3 mL of hydrochloric acid. Proceed as directed in the test solution, then add the volume of standard lead solution directed in the monograph and water to make 50 mL.

7.4   COMMON PROBLEMS AND SOLUTIONS

1. How should the criteria for selections from methods 1 to 4 of the test be established, and the solution preparation specified in the heavy metals limit test in JP be controlled?

- Method 1 can be used if the drug substance can be dissolved in 40 mL of water. There is no precipitation after adding 2 mL of dilute acetic acid. The pH of the solution can easily be adjusted to 3.0 to 3.5. It results in a favorable recovery rate.
- Method 2 is commonly used, while method 3 is rarely used for items listed in JP. Method 4 should be used when an unsatisfactory recovery rate is obtained with method 2.

2. The spiked recovery test cannot produce a favorable result. The following precautions should be taken at the time of preparation of the test and control solutions from samples.

- Carbonize by gradual ignition to avoid burning of the sample or loss by swelling outside the crucible.
- For incineration with sulfuric acid in methods 2 and 4, be sure to limit the temperature to 500 to 600°C to avoid volatilization of heavy metals.
- Ensure that the pH of the test solution is in the range 3.0 to 4.0 using dilute acetic acid during preparation.
- In the case of method 4, use a slightly larger crucible to prevent scattering of the sample during burning with ethanol.

3. After adding sodium sulfide TS, turbidity is observed when the test and control solutions are compared.

- It is recommended to add sodium sulfide reagent solution to the test and control solutions at the same time and to observe them approximately 5 min later. After that time, sulfur will precipitate causing turbidity.

4. Even after addition of sodium sulfide TS and mixing well, there is only slight darkening, or there is existing coloring in the reagent solution.

- Sodium sulfide TS contains glycerin to stabilize and inhibit oxidation. The ability to produce sulfide is decreased with storage time. It is recommended that the fresh reagent solution should be used, especially for the quantitation test. Even if the reagent solution is refrigerated, it should be used within a couple of months.

REFERENCES

1. *Japanese Pharmacopoeia*, JP XIV, General Tests, Processes and Apparatus, 26, Heavy Metals Limit Test, p. 43. See *http://jpdb.nihs.go.jp/jp14e/contents.html*.
2. ICH Harmonized Guidelines, ICH T Q2A, *Text on Validation of Analytical Methods*, Mar. 1995; ICH Q2B, *Validation of Analytical Procedures: Methodology*, May 1997. See *www.nihs.go.jp/dig/ich/ichindex.htm*.

8

BIOANALYTICAL METHOD VALIDATION

Fabio Garofolo, Ph.D.
Vicuron Pharmaceuticals, Inc.

8.1 INTRODUCTION

8.1.1 Definition of Bioanalytical Method Validation

Bioanalytical method validation is a procedure employed to demonstrate that an analytical method used for quantification of analytes in a biological matrix is reliable and reproducible to achieve its purpose: to quantify the analyte with a degree of accuracy and precision appropriate to the task. Validation data, through specific laboratory investigations, demonstrate that the performance of a method is suitable and reliable for the analytical applications intended. The quantitative approach used in bioanalytical methods involves the use of a standard curve method with internal standard. In this approach the analyte concentration can be assigned by referring the response to other samples, called *calibrators* or *calibration standards*. In addition to the samples of unknown concentration, the bioanalytical set includes the calibration standards, and samples containing no analyte, called *blanks*, to assure that there are no interferences in the matrix. Accuracy and precision of the method are calculated using the back-calculated concentrations of samples of known composition called *quality control samples* (QCs). The calibrator standards and quality control samples should be prepared in the same matrix as the actual samples.

All these checks should be performed to guarantee the reliability of selective and sensitive bioanalytical method before applying them for the quantitative

Analytical Method Validation and Instrument Performance Verification, Edited by Chung Chow Chan, Herman Lam, Y. C. Lee, and Xue-Ming Zhang
ISBN 0-471-25953-5 Copyright © 2004 John Wiley & Sons, Inc.

evaluation of drugs and their metabolites. The data generated by these methods must be very reproducible and reliable because they are used in the evaluation and interpretation of bioavailability, bioequivalence, pharmacokinetic, and preclinical findings [13,15,18]. A validation is required for all the bioanalytical methods employing analytical techniques such as gas chromatography (GC); high-pressure liquid chromatography (HPLC); hyphenated mass spectrometric techniques such as LC-MS, LC-MS/MS, GC-MS, and GC-MS/MS; or immunological and microbiological procedures such as radioimmunoassay (RIA) and enzyme-linked immunosorbant assay (ELISA) for the quantitative determination of drugs and/or metabolites in biological matrices such as blood, serum, plasma, urine, feces, saliva, sputum, cerebrospinal fluid (CSF), tissues, and skin samples.

Although there are various stages in the validation of a bioanalytical procedure, the process by which a specific assay is developed, validated, and used in routine sample analysis can be divided into four main steps:

1. Reference standards preparation (stock solutions, working solutions, spiked calibrators, and QCs) (see Section 8.2.1)
2. Bioanalytical method development where the assay procedure is established
3. Bioanalytical method validation and definition of the acceptance criteria for the analytical run and/or batch
4. Application of validated method to routine sample analysis

The fundamental parameters for bioanalytical validations include accuracy, precision, selectivity, sensitivity, reproducibility, stability of the drug in the matrix under study storage conditions, range, recovery, and response function (see Section 8.2.1). These parameters are also applicable to microbiological and ligand-binding assays. However, these assays possess some unique characteristics that should be considered during method validation, such as selectivity and quantification issues.

8.1.2   Regulatory Guidance on Bioanalytical Method Validation

During the 1990 Washington Conference on Analytical Methods Validation: Bioavailability, Bioequivalence and Pharmacokinetic Studies [1], parameters that should be used for method validation were defined. The final report of this conference is considered the most comprehensive document on the validation of bioanalytical methods. Many multinational pharmaceutical companies and contract research organizations contributed to its final draft. This scientific meeting was sponsored by the American Association of Pharmaceutical Scientists (AAPS), the Association of Official Analytical Chemists (AOAC), and the U.S. Food and Drug Administration (FDA). The conference report has been used as a reference by bioanalytical laboratories and regulatory agencies worldwide.

However, in the last decade there has been a tremendous advance in the techniques used for sample preparation and analysis. For instance, in the mass spectrometry field, new interfaces, ionization, and detection techniques were developed. This technical advancement leads to the use of commercial hyphenated mass spectrometric techniques and automation as preferred instrumentation for bioanalytical methods. LC-MS/MS assays replaced the conventional LC and GC, and the use of multiwell plates, automated robotic sample processing (Multiprobe, Tomtec), and online extraction techniques (Prospeck, Turbulent Flow Chromatography) took over manual sample preparation procedures.

The worldwide use of these new powerful bioanalytical techniques, characterized by more rapid throughput and increase in sensitivity [12,16] required a review of the original 1990 Washington report. The Guidelines of the 1990 Conference were initially reviewed during a meeting in June 1994 in Munich, Germany [2]. This meeting focused primarily on the critical and statistical evaluation of the acceptance criteria defined in the guidelines and on formulating recommendations to improve the guidelines.

A conference titled "Bioanalytical Methods Validation: A Revisit with a Decade of Progress" was held on January 2000, again in Washington, DC. The objective of this conference was to reach a consensus on what should be required in bioanalytical methods validation, and which procedures should be used to perform the validation [3]. The FDA 2001 Guidance for Industry on Bioanalytical Method Validation [4] is based on the final report of both the 1990 and 2000 Washington conferences. At the beginning of this document the FDA states very clearly that its guidance for bioanalytical method validation represents its current thinking on this topic and that an alternative approach may be used if such an approach satisfies the requirements of applicable statutes and regulations [4]. This statement allows bioanalytical laboratories to adjust or modify the FDA recommendations, depending on the specific type of bioanalytical method used.

Compliance with the FDA guidance can be considered a minimum requirement to test the performance of a bioanalytical method. Due to the fact that the validation process should simulate closely sample analysis, the real and decisive final test for a "validated" method will always be the sample analysis itself. It is possible that even if it passes all the validation criteria, a bioanalytical method may not be reliable for the analysis of actual samples. This undesirable situation could happen when actual samples (in vivo samples) contain new interferences not present in the spiked samples (in vitro samples) due to a metabolic process and/or other biotransformations. For this reason, bioanalytical laboratories could decide to use more stringent criteria and procedures and/or use actual sample during the method development to further guarantee the validity of the validated methods.

In the following section we summarize the current general recommendations for bioanalytical method validation practices according to the FDA guidelines, with other alternative approaches to be discussed later.

8.2   CURRENT VALIDATION PRACTICE

The validation procedures for bioanalytical methods are in continuous evolution since bioanalytical methods are constantly undergoing changes in improvements, and in many instances they are at the cutting edge of the technology. An overview of the FDA Guidance for Industry, Bioanalytical Methods Validation, May 2001 [4], is reported here as a reference for current validation practice.

8.2.1 Definitions

As the first step in understanding the procedure used for the validation of bioanalytical methods, it is important to have clearly in mind definitions of the analytical terms used.

- *Accuracy:* the degree of closeness of the determined value to the nominal or known true value under prescribed conditions. This is sometimes termed *trueness.*
- *Analyte:* a specific chemical moiety being measured, which can be intact drug, biomolecule, or its derivative, metabolite, and/or degradation product in a biological matrix.
- *Analytical run (or batch):* a complete set of analytical and study samples with the appropriate number of standards and QCs for their validation. Several runs (or batches) may be completed in one day, or one run (or batch) may take several days to complete.
- *Biological matrix:* a discrete material of biological origin that can be sampled and processed in a reproducible manner. Examples are blood, serum, plasma, urine, feces, saliva, sputum, and various discrete tissues.
- *Stock solutions:* the original solutions prepared directly by weighing the reference standard of the analyte and dissolving it in the appropriate solvent. Usually, stock solutions are prepared at a concentration of 1 mg/mL in methanol and kept refrigerated at $-20°C$ if there are no problems of stability or solubility.
- *Calibration standard:* a biological matrix to which a known amount of analyte has been added or *spiked*. Calibration standards are used to construct calibration curves from which the concentrations of analytes in QCs and in unknown study samples are determined.
- *Internal standard:* test compound(s) (e.g., structurally similar analog, stable labeled compound) added to both calibration standards and samples at known and constant concentration to facilitate quantification of the target analyte(s).
- *Limit of detection (LOD):* the lowest concentration of an analyte that the bioanalytical procedure can reliably differentiate from background noise.
- *Lower limit of quantification (LLOQ):* the lowest amount of an analyte in a sample that can be determined quantitatively with suitable precision and accuracy.
- *Matrix effect:* the direct or indirect alteration or interference in response due to the presence of unintended analytes (for analysis) or other interfering substances in the sample.

- *Method:* a comprehensive description of all procedures used in sample analysis.
- *Precision:* the closeness of agreement (*degree of scatter*) between a series of measurements obtained from multiple sampling of the same homogeneous sample under the prescribed conditions.
- *Processed sample:* the final extract (prior to instrumental analysis) of a sample that has been subjected to various manipulations (e.g., extraction, dilution, concentration).
- *Quantification range:* the range of concentration, including ULOQ and LLOQ, that can be quantified reliably and reproducibly with accuracy and precision through the use of a concentration–response relationship.
- *Recovery:* the extraction efficiency of an analytical process, reported as a percentage of the known amount of an analyte carried through the sample extraction and processing steps of the method.
- *Reproducibility:* the precision between two laboratories. It also represents precision of the method under the same operating conditions over a short period of time.
- *Sample:* a generic term encompassing controls, blanks, unknowns, and processed samples, as described below:
  - *Blank:* a sample of a biological matrix to which no analytes have been added that is used to assess the specificity of the bioanalytical method.
  - *Quality control sample (QC):* A spiked sample used to monitor the performance of a bioanalytical method and to assess the integrity and validity of the results of the unknown samples analyzed in an individual batch.
  - *Unknown:* a biological sample that is the subject of the analysis.
- *Selectivity:* the ability of the bioanalytical method to measure and differentiate the analytes in the presence of components that may be expected to be present. These could include metabolites, impurities, degradants, or matrix components.
- *Stability:* the chemical stability of an analyte in a given matrix under specific conditions for given time intervals.
- *Standard curve:* the relationship between the experimental response value and the analytical concentration (also called a *calibration curve*).
- *System suitability:* determination of instrument performance (e.g., sensitivity and chromatographic retention) by analysis of a reference standard prior to running the analytical batch.
- *Upper limit of quantification (ULOQ):* the highest amount of an analyte in a sample that can be determined quantitatively with precision and accuracy.
- *Validation*
  - *Full validation:* establishment of all validation parameters to apply to sample analysis for the bioanalytical method for each analyte.
  - *Partial validation:* modification of validated bioanalytical methods that do not necessarily call for full revalidation.

- *Cross-validation:* comparison of validation parameters of two bioanalytical methods.
- *Working solutions:* solutions prepared from the stock solution through dilution in the appropriate solvent at the concentration requested for spiking the biological matrix.

8.2.2 Selectivity

For selectivity, there should be evidence that the substance being quantified is the intended analyte. Therefore, analyses of blank samples of the appropriate biological matrix (plasma, urine, or other matrix) should be obtained from *at least six sources.* Each blank sample should be tested for interference, and selectivity should be ensured at the lower limit of quantification (LLOQ).

Potential interfering substances in a biological matrix include endogenous matrix components, metabolites, decomposition products, and in the actual study, concomitant medication. Whenever possible, the *same biological matrix* as the matrix in the intended samples should be used for validation purposes. For tissues of limited availability, such as bone marrow, physiologically appropriate proxy matrices can be substituted. Method selectivity should be evaluated during method development and method validation and can continue during the analysis of actual study samples.

As with chromatographic methods, *microbiological and ligand-binding assays* should be shown to be selective for the analyte. The following recommendations for dealing with two selectivity issues should be considered:

1. Specific interference from substances physiochemically similar to the analyte:
   a. Cross-reactivity of metabolites, concomitant medications, or endogenous compounds should be evaluated individually and in combination with the analyte of interest.
   b. Whenever possible, the immunoassay should be compared with a validated reference method (such as LC-MS) using incurred samples and predetermined criteria for agreement of accuracy of immunoassay and reference method.
   c. The dilutional linearity to the reference standard should be assessed using study (incurred) samples.
   d. Selectivity may be improved for some analytes by incorporation of separation steps prior to immunoassay.
2. Nonspecific matrix effects:
   a. The standard curve in biological fluids should be compared with standard in buffer to detect matrix effects.
   b. Parallelism of diluted study samples should be evaluated with diluted standards to detect matrix effects.
   c. Nonspecific binding should be determined.

8.2.3 Reference Standard

The purity of the reference standard used to prepare spiked samples can affect the study data. For this reason, an *analytical reference standard* of known identity and purity should be used to prepare solutions of known concentrations. If possible, the reference standard should be identical to the analyte. When this is not possible, an established chemical form (free base or acid, salt, or ester) of known purity can be used. Three types of reference standards are generally used:

1. Certified reference standards (e.g., USP compendial standards)
2. Commercially supplied reference standards obtained from a reputable commercial source
3. Other materials of documented purity, custom-synthesized by an analytical laboratory or other noncommercial establishment

The source and lot number, expiration date, certificates of analyses when available, and/or internally or externally generated evidence of identity and purity should be furnished for each reference standard.

8.2.4 Standard Curve

A calibration curve should be generated for each analyte in the sample. A *sufficient number of standards* should be used to adequately define the relationship between concentration and response. A calibration curve should be prepared in the same biological matrix as the samples in the intended study by spiking the matrix with known concentrations of the analyte. The number of standards used in constructing a calibration curve will be a function of the anticipated range of analytical values and the nature of the analyte–response relationship (detector sensitivity and detector linear range). Concentrations of standards should be chosen based on the concentration range expected in a particular study. A calibration curve should consist of a blank sample (a matrix sample processed without an internal standard), a zero matrix sample processed with an internal standard, and six to eight nonzero samples covering the range expected, including LLOQ (see Section 8.2.1).

Lower Limit of Quantification. The lowest standard on the calibration curve should be accepted as the limit of quantification if the following conditions are met:

1. The analyte response at the LLOQ should be *at least five times* the blank response.
2. Analyte peak (response) should be identifiable, discrete, and reproducible, with a *precision of less than 20% coefficient of variation (CV) and an accuracy of 80 to 120%.*

Concentration–Response Relationship. The *simplest model* that describes the concentration–response relationship adequately should be used. The selection of weighing and use of a complex regression equation should be *justified.*

The following conditions should be met in developing a calibration curve:

1. Less than 20% deviation of the LLOQ from nominal concentration
2. 15% deviation of standards other than LLOQ from nominal concentration

At least four of six nonzero standards should meet the criteria above, including the LLOQ and the calibration standard at the highest concentration. In addition, excluding the standards that did not meet the criteria should not change the concentration response model used. A sufficient number of standards should be used to adequately define the relationship between concentration and response. More standard concentrations may be recommended for nonlinear than for linear relationships.

Microbiological and Immunoassay. Microbiological and immunoassay standard curves are inherently *nonlinear*, and in general, more concentration points may be recommended to define the fit over the standard curve range than for chemical assays. In addition to their nonlinear characteristics, the response–error relationship for immunoassay standard curves is a variable function of the mean response (heteroscedisticity). For these reasons, a minimum of six nonzero calibrator concentrations in duplicate is recommended. The concentration–response relationship is most often fitted to a four- or five-parameter logistic model, although others may be used with suitable validation. The use of anchoring points in the asymptotic high- and low-concentration ends of the standard curve may improve the overall curve fit. Generally, these anchoring points will be at concentrations that are below the established LLOQ and above the established ULOQ. In the case where replicate samples should be measured during the validation to improve accuracy, the same procedure should be followed as for unknown samples.

8.2.5   Precision and Accuracy

Accuracy is determined by replicate analysis of samples containing known amounts of the analyte. Accuracy should be measured using a *minimum of five determinations* per concentration. A *minimum of three concentrations* in the range of expected concentrations is recommended. The mean value should be within 15% of the actual value except at LLOQ, where it should not deviate by more than 20%. The deviation of the mean from the true value serves as the measure of accuracy.

Precision should be measured using a *minimum of five determinations* per concentration. A *minimum of three concentrations* in the range of expected concentrations is recommended. The precision determined at each concentration level should not exceed 15% of the CV except for the LLOQ, where it should not exceed 20% of the CV. Precision is further subdivided into:

- Within-run, intrabatch precision or repeatability, which assesses precision during a single analytical run
- Between-run, interbatch precision or intermediate precision, which measures precision with time and may involve different analysts, equipment, reagents, and laboratories

In consideration of *high throughput analyses*, including but not limited to multiplexing, multicolumn, and parallel systems, sufficient QC samples should be used to ensure control of the assay. The number of QC samples to ensure proper control of the assay should be determined based on the run size. The placement of QC samples should be considered judiciously in the run. At a minimum, three concentrations representing the entire range of the standard curve should be studied: one within three times the lower limit of quantification (LLOQ) (low QC sample), one near the center (middle QC), and one near the upper boundary of the standard curve (high QC). Reported method validation data and the determination of accuracy and precision should include all outliers; however, calculations of accuracy and precision, excluding values that are statistically determined as outliers, can also be reported.

8.2.6 Dilutions

The ability to dilute samples originally above the upper limit of the standard curve should be demonstrated by accuracy and precision parameters in the validation.

8.2.7 Recovery

Recovery of the analyte need not be 100%, but the extent of recovery of an analyte and of the internal standard should be *consistent, precise*, and *reproducible*. Recovery experiments should be performed by comparing the analytical results for extracted samples at three concentrations (low, medium, and high) with unextracted standards that represent 100% recovery.

It may be important to consider the variability of the matrix due to the physiological nature of the sample. In the case of LC-M/MS-based procedures, appropriate steps should be taken to ensure the lack of matrix effects throughout application of the method, especially if the nature of the matrix changes from the matrix used during method validation. For *Microbiological and immunoassay*, if separation is used prior to assay for study samples but not for standards, it is important to establish recovery and use it in determining results. In this case, possible approaches to assess efficiency and reproducibility of recovery are:

- The use of radiolabeled tracer analyte (quantity too small to affect the assay)
- The advance establishment of reproducible recovery
- The use of an internal standard that is not recognized by the antibody but can be measured by another technique

8.2.8 Stability

The stability of an analyte in a particular matrix and container system is relevant only to that matrix and container system and *should not be extrapolated* to other matrices and container systems. Stability procedures should evaluate the stability of the analytes during sample collection and handling, after long-term (frozen at the intended storage temperature) and short-term (benchtop, room-temperature) storage, and after going through freeze–thaw cycles and the analytical process. Conditions used in stability experiments should reflect situations likely to be encountered during actual sample handling and analysis. The procedure should include an evaluation of analyte stability in stock solution. For compounds with potentially labile metabolites, the stability of analyte in matrix from dosed subjects (or species) should be confirmed. All stability determinations should use a set of samples prepared from a freshly made stock solution of the analyte in the appropriate analyte-free, interference-free biological matrix.

Freeze and Thaw Stability. Analyte stability should be determined after three freeze–thaw cycles. At least three aliquots at each of the low and high concentrations should be stored at the intended storage temperature for 24 h and thawed unassisted at room temperature. When thawed completely, the samples should be refrozen for 12 to 24 h under the same conditions. The freeze–thaw cycle should be repeated twice more, then analyzed on the third cycle. If an analyte is unstable at the intended storage temperature, the stability sample should be frozen at $-70°C$ during the three freeze–thaw cycles.

Short-Term Temperature Stability. Three aliquots of each of the low and high concentrations should be thawed at room temperature and kept at this temperature from 4 to 24 h (based on the expected duration that samples will be maintained at room temperature in the intended study) and analyzed.

Long-Term Stability. The storage time in a long-term stability evaluation should *exceed* the time between the date of first sample collection and the date of last sample analysis. Long-term stability should be determined by storing at least three aliquots of each of the low and high concentrations under the same conditions as the study samples. The volume of samples should be sufficient for analysis on three separate occasions. The concentrations of all the stability samples should be compared to the mean of back-calculated values for the standards at the appropriate concentrations from the first day of long-term stability testing.

Stock Solution Stability. The stability of stock solutions of drug and the internal standard should be evaluated at room temperature for at least 6 h. If the stock solutions are refrigerated or frozen for the relevant period, the stability should be documented. After completion of the desired storage time, the stability should be tested by comparing the instrument response with that of freshly prepared solutions.

Postpreparative Stability. The stability of processed samples, including the resident time in the autosampler, should be determined. The stability of the drug and the internal standard should be assessed over the anticipated run time for the batch size in validation samples by determining concentrations on the basis of original calibration standards. Reinjection reproducibility should be evaluated to determine if an analytical run could be reanalyzed in the case of instrument failure.

8.2.9   Additional Verification

During the course of a typical drug development program, a bioanalytical method will undergo many modifications. These evolutionary changes (e.g., addition of a metabolite, lowering of the LLOQ,) are required to support specific studies and will require a different level of validation to demonstrate the validity of assay performance. These modifications should be validated to continue to ensure suitable performance of the method. When changes are made in a previously validated method, the analyst should exercise judgment as to how much additional validation is needed. Various types and levels of validation are defined and characterized below.

Full Validation. Full validation is important when developing and implementing a bioanalytical method for the first time, for a new drug entity, and if metabolites are added to an existing assay for quantification.

Partial Validation. Partial validations are modifications of already validated bioanalytical methods. Partial validation can range from as little as one intraassay accuracy and precision determination to a nearly full validation. Typical bioanalytical method changes that fall into this category include but are not limited to:

- Bioanalytical method transfers between laboratories or analysts
- Change in analytical methodology (e.g., change in detection systems)
- Change in anticoagulant in harvesting biological fluid
- Change in matrix within species (e.g., human plasma to human urine)
- Change in sample processing procedures
- Change in species within matrix (e.g., rat plasma to mouse plasma)
- Change in relevant concentration range
- Changes in instruments and/or software platforms
- Limited sample volume (e.g., pediatric study)
- Rare matrices
- Selectivity demonstration of an analyte in the presence of concomitant medications
- Selectivity demonstration of an analyte in the presence of specific metabolites

Cross-Validation. Cross-validation is a comparison of validation parameters when two or more bioanalytical methods are used to generate data within the same study or across different studies. An example of cross-validation would be a situation where an original validated bioanalytical method serves as the *reference* and the revised bioanalytical method is the *comparator*. The comparisons should be done both ways. When sample analyses within a single study are conducted at more than one site or more than one laboratory, cross-validation with spiked matrix standards and subject samples should be conducted at each site or laboratory to establish interlaboratory reliability. All modifications should be assessed to determine the degree of validation recommended. Immunoassay reoptimization or validation may be important when there are changes in key reagents.

8.2.10   Documentation

A specific, detailed description of the bioanalytical method should be written. This can be in the form of a protocol, study plan, report, and/or standard operating procedure (SOP). All experiments used to make claims or draw conclusions about the validity of the method should be presented in a report (method validation report).

8.3   COMMON PROBLEMS AND SOLUTIONS

The 2000 Washington Conference on Bioanalytical Method Validation [3] reviewed the progress, impact, and advances made during the last decade of bioanalytical methods validation since the first Washington conference in 1990. Hyphenated mass spectrometric–based assays, ligand-based assay, and high-throughput systems were discussed in depth. However, there are still some controversies on some scientific approaches and criteria used during the validation process. Some of the most interesting issues discussed during the last 10 years are discussed in the following paragraphs.

8.3.1   Definitions

The glossaries in the 1990 and 2000 Washington conference final reports [1,3] define most of the analytical terms used in the validation of a method. However, internationally accepted definitions such as those by ISO or IUPAC already exist and have been elaborated carefully over many years [2,6]. The definitions reported in the 1990 and 2000 Washington conference final reports sometime agree only partially with the ISO and IUPAC. Following are some examples for comparison.

Limit of Detection
2000 Conference: the lowest concentration of an analyte that a bioanalytical procedure can reliably differentiate from background noise.

1990 Conference: the lowest concentration of an analyte that an analytical process can reliably differentiate from background levels.

United States Pharmacopoeia: "thus, limit tests merely substantiate that the analyte concentration is above or below a certain level. ..."

IUPAC: the concentration giving a signal three times the standard deviation of the blank.

Lower Limit of Quantification

2000 Conference: the lowest amount of an analyte in a sample that can be determined quantitatively with suitable precision and accuracy.
  Comment:

- This definition is connected with the definition of sensitivity of the method as the concentration of the lowest standard with a coefficient of variance (CV) $\leq 20\%$.

1990 Conference: the lowest concentration of an analyte that can be measured with a stated level of confidence.

IUPAC: the concentration that gives rise to a signal 10 times the standard deviation of the blank

Upper Limit of Quantification

2000 Conference: the highest amount of an analyte in a sample that can be determined quantitatively with precision and accuracy.
  Comment:

- From a practical point of view this definition can be interpreted as being imposed by the linear boundary of the calibration curve (quadratic behavior) due to saturation of the detector or/and ion suppression effect and/or contamination for low-level samples (carryover) (see Section 8.3.7).

Accuracy

2000 Conference: the degree of closeness of the determined value to the nominal or known true value under prescribed conditions. This is sometimes termed *trueness*.
  Comments:

- This is expressed as percent relative error (% RE).
- RE may be positive, negative, or zero.
- % RE = [(mean value/theoretical value) − 1] × 100.
- In general, the measured concentration of sample of known composition compared to the theoretical concentration over an appropriate range of concentration is considered an indication of accuracy.
- Intraassay accuracy is the RE of the mean of the replicate analysis of a validation sample during a single validation batch.
- Interassay accuracy is the RE of the overall mean of the replicate analyses of a validation sample for all validation batches.

1990 Conference: the closeness of the determined value to the true value. Generally, recovery of added analyte over an appropriate range of concentrations is taken as an indication of accuracy. Whenever possible, the concentration range chosen should bracket the concentration of interest.

ISO: the closeness of agreement between the test result and the accepted reference value.
Comment:

- The definition of accuracy reported in the 1990 Washington conference glossary was partially reformulated in the 2000 conference. The first sentence of both definitions is close to the ISO definition. Some disagreements were raised [2] on the second sentence of the 1990 definition during the 1994 meeting in Germany: "It is correct that recovery can be taken as an indication that a method is accurate but it is no more than an indication. Inclusion of recovery in a definition of accuracy may lead some analysis to conclude that adequate recovery always means that a method is accurate and that, of course, is not true. Suppose that a method is not selective and that some interference is also measured. The result will then be a certain (approximately the same) amount too high in both the unspiked and the spiked sample. However, the difference between the two results from which the recovery is calculated, will be correct, leading to the false conclusion that the method is accurate."

8.3.2 Selectivity/Specificity

Selectivity is the ability of the bioanalytical method to measure and differentiate the analytes in the presence of components that may be expected to be present. *Specificity* is the ability to assess unequivocally the analyte in the presence of components that may be expected to be present. In general, analytical methods are selective, and only in same cases also specific (e.g., an LC-MS/MS bioanalytical method is highly selective but not always also specific because it could be possible to find in the complex biological matrix an interference with the same retention time, molecular weight, and main fragment of the analyte of interest). Even if the 2000 Washington conference focuses only on selectivity, it is up to bioanalytical laboratories to differentiate in their documentation between selectivity and specificity or consider them equivalent and use them interchangeably.
A general approach to prove the selectivity (specificity) of the method is to verify that:

- The response of interfering peaks at the retention time of the analyte is less than 20% of the response of an LLOQ standard, or the response at the LLOQ concentration is at least five times greater than any interference in blanks at the retention time of the analyte.
- The responses of interfering peaks at the retention time of the internal standard are ≤5% of the response of the concentration of the internal standard used in the studies.

The 2000 Washington conference guidelines suggest analyzing at least six independent sources (i.e., six different lots or six different individual samples) of blank plasma or urine separately, and demonstrate that no substances are present that interfere with the quantification of the analyte. The degree of compliance with this guideline changes from lab to lab as the number of independent sources tested.

In general, from six to a maximum of 10 independent sources of blank matrix could be chosen to be tested. All six to 10 blanks should pass the selectivity acceptance criteria.

- *Case A:* If the out of six to 10 does not pass the criteria, the method must go back to the development step, and the method selectivity should be improved (improvement of the cleanup/extraction step and/or chromatographic separation).
- *Case B:* Even if a maximum of the out of six to 10 lots of matrix may not pass the criteria, the overall method can still be considered selective and does not need further development:
  - *Case B1:* If two out of six to 10 do not pass the criteria, the method must go back to the development step and the method selectivity should be improved.
  - *Case B2:* If two out of six to 10 lots of blank matrix may not pass the criteria another six to 10 additional lots may be screened and all the new additional lots must pass the criteria, or additional development is necessary to eliminate the source of interference.
  - *Case B3:* If two out of six to 10 lots of blank matrix may not pass the criteria another six to 10 additional lots may be screened, and if two or more of the new additional lots do not pass the criteria, the method must go back to the development step and the method selectivity should be improved.

It would be worthwhile to consider that evaluation of the interference is performed through quantification of the interfering compound at a level below the LLOQ (criteria $\leq 20\%$, i.e., outside the validated quantification range). Outside this range the accuracy of the results produced is unknown. The same considerations can be used for evaluation of the method carryover (see Section 8.3.7).

Limited Quantity of Matrices. For biological matrices that are difficult to obtain, only a single source may be available and is therefore the only possibility for testing selectivity. These matrices include, but are not limited to, human and primate tissue and cerebrospinal fluid. In the absence of the corrected biological matrix, as the 2000 Washington conference final report suggests, a physiologically appropriate proxy matrix can be used.

Microbiological and Ligand Binding. For microbiological and ligand-binding assays [1,3,4], an appropriate combination of other techniques may be used to show selectivity, including the following:

- Standards in biological fluids are compared with standards in buffer to detect matrix effects.
- Parallelism of diluted clinical samples with diluted standards is used to detect the presence of closely related compounds.
- Serial separation techniques (e.g., extraction, chromatography), with the bioassay as detector are used to demonstrate that the response is due only to the analyte in question.
- Metabolite (or endogenous compound) cross-reaction may be initially assessed by comparison of displacement curves but in critical cases should also be assessed by addition of metabolite to analyte.
- Similar criteria will be applicable when the drug is administered concomitantly with other drugs.

8.3.3   Reference Standard

"The quantitative analysis is only as good as the reference material allows." The reference standard is used as the primary standard against which all determinations are made. Reference standard characteristics should be appropriately determined and documented for each batch. Copies of this documentation must be included with study records and must be available for inspection. The reference standard should be (1) of the highest purity obtainable, (2) independently characterized to establish identity, strength or potency, and purity, (3) in the most stable form, and (4) stored under suitable conditions.

This information should be included in documentation such as the certificate of analysis (CoA), test article characterization (TAC), and reference standard profile. Certified reference standards can be purchased from appropriate suppliers. If standards are not available, the recommendation is to collect or synthesize enough material, and analyze, certify, and use it as the standard. Following are some considerations:

1. All the concentration should be corrected for potency and should be expressed either as the free acid or the free base, as appropriate.
2. If the reference material is a salt or contains water of crystallization, this must be taken into account in all the calculations.
3. If the potency is later determined to be different from what was used during the method validation, the impact of this new information must be evaluated and documented. A partial validation could be necessary.
4. To start the validation the documentation of the potency of the main analytes must be available( ). In some bioanalytical laboratories this documentation is not always required for analytes that are metabolites and/or internal standards.
5. As far as the preparation of the stock solutions for calibration standards and quality control samples is concerned, different labs use different approaches. In general, the calibrators used for the validation should be prepared from

an analyte weighing, which is separated from the weighting used to prepare the quality control samples. If the quantity of the reference standard
is not sufficient for separate weighings for calibrators and quality control
samples, only one weighing could be used, but this will be documented
clearly in the final report.

Reagents and Solutions. As for good laboratory practice, all reagents and solutions in the laboratory areas should be labeled to indicate identity, title or concentration, storage requirements, and expiration date. Deteriorated or outdated
reagents and solutions must not be used. Each storage container for a test or control article should be labeled by name, chemical abstract number, code number,
batch number, expiration date, storage conditions (if any, and, where appropriate),
strength, purity, and composition.

8.3.4   Standard Curve

When the standard curve has been established and the LLOQ and ULOQ validated, the assessment of unknown concentrations by extrapolation is not allowed
beyond the validated range. The most accurate and precise estimates of concentration is in the linear portion of the curve even if acceptable quantitative results
can be obtained up to the boundary of the curve using a quadratic model. For
a linear model, statistic calculations suggest a minimum of six concentrations
evenly placed along the entire range assayed in duplicate [5,7,8].

As for acceptance criteria, the 2000 Washington conference guidelines for
the linear model no longer request that the correlation coefficient (r) be 0.95 or
greater. The guidelines use the $\pm20/15$ criteria (i.e., the accuracy of the calibration
standards should be within $\pm20/15$). A model for regression should be chosen during method development when the performance of calibration standards
is studied, and the regression chosen must not be changed during validation,
or worse, during sample analysis. Standard curve fitting should be determined
by applying the simplest algorithm (model) that best describes the concentration–response relationship using appropriate weighing and statistical tests for
goodness-of-fit requirements.

In the regression model, a polynomial function is considered: $y = b_0 + b_1x +
b_2x^2 + b_3x^3 + \cdots$; for the linear model, the terms x^2 and larger are ignored;
and for the quadratic model, terms larger than x^2 are not considered. Even if
the use of the quadratic model is allowed and used extensively by some bioanalytical laboratories, the use of linear regression models should be attempted
first. Usually, a deviation from the linear model should be investigated and
avoided. Nonlinearity could be due to (1) injection techniques, (2) sample holdup
on glassware, (3) cross-talk in MS/MS, (4) interferences, and (5) too wide a concentration range.

The 2000 Washington conference final report stated clearly (see Section 8.2.4)
that selection of weighing should be justified. In general, it is possible to justify
the weighing used considering that the data produced by bioanalytical methods do

not meet the homoscedasticity assumption (condition of equal variance). There-fore, the higher concentrations have a greater influence on the fitted line when least squares linear regression is used.

Weighing functions reduce the influence of values obtained for higher concentrations on slope and intercept. From statistical considerations the most common weighing function for LC-MS- and LC-MS/MS-based assays should be $1/x$, due to the fact that variance in y increases in proportion to the concentration [5]. In HPLC-UV-based assays the use of $1/x^2$ weighed linear regression analysis can significantly reduce the LLOQ obtainable, where the standard deviation of y varies with x [7,8]. A comparison between the weighed least squares procedure and the conventional least squares calibration shows improvements in accuracy at the lower end. The principal advantage in this case is for clinical pharmacology and pharmacokinetic studies when concentration values being measured by the method are near LLOQ.

Rejection of Calibration Standard. A calibration standard should be eliminated from the calculations if (1) a problem has been discovered and documented during the sample processing or (2) a chromatographic problem or system mal-function has been discovered and documented. If the back-calculated value is $>20\%$ higher or lower than the theoretical value for LLOQ or $>15\%$ higher or lower than the theoretical value for all other standards, two different approaches can be taken:

1. The simplest and strict approach, used by many laboratories, is to eliminate all the calibration standards that do not fall within $\pm 15\%$ of theoretical (or $\pm 20\%$ of theoretical for the LLOQ).
2. A second approach is to define if the calibration standard falling outside the acceptance criteria is or is not an outlier. All the calibration standards identified statistically as outliers must be rejected and not used in the standard curve regression.

An outlier is an individual measurement that appears to be statistically different from others when a series of replicate measurements is performed. The Q test is an easy test used to define an outlier.

$$Q_{observed} = \frac{\text{difference between the questionable data point and the nearest value}}{\text{difference between the lowest and highest values in the curve}}$$

If the $Q_{observed} > Q_{theoretical}$, the outlier may be discarded. The Q test gives a good evaluation of the outlier when the number of observations ≥ 5. Different intervals of confidence can be chosen to evaluate $Q_{theoretical}$ even if a 95% confidence limit is suggested.

Another approach is to consider the statistical calculation based on a Student's t distribution assumption with 95% probability (t_{95}). If the ratio of relative error

(RE) of the questionable calibrator with the standard deviation (SD) of all calibrators (RE/SD) is greater than the absolute t_{95} value at the corresponding degree of freedom (number of independent calibrators minus 1), the individual calibrator is identified as an outlier (i.e., if $RE/SD > t_{95}$, the outlier may be discarded, where RE is the difference between the percent accuracy value of an individual standard and the mean of the accuracy values of n independent standards and SD is the closeness of the replicate measurements in a set, i.e., the spread of data around the mean).

There are other statistical approaches that can be used to detect and eliminate outliers. In summary, an outlier should be rejected if it is outside the 95% confidence limit of regression line.

8.3.5 Precision and Accuracy

The magnitude of the uncertainty associated with a measurement should always be evaluated even if method development is able to generate the best, error-free standard curve possible in order to obtain the best, error-free concentration values for the unknown. Precision and accuracy values define the compromise between the demand for certainty in reported results and the inherent uncertainty in bioanalytical measurement. Precision and Accuracy are the most important parameters used to set the performance standards of bioanalytical methods. They define if the zone of uncertainty connected with the data produced by the method is or is not acceptable for the specific task [13,21].

Systematic Error and Random Error. An analytical result can be affected by a combination of two different kinds of experimental error: systematic error and random error. *Systematic errors* are associated with the accuracy of an analytical method and are the difference of the mean value from the true value. The measure of this difference is the bias. The mean is only an estimate of the true value because only a limited number of experiments can be carried out. The estimate becomes better when more experiments have been carried out. In conclusion, since all the measurements are estimates, the true value can be approached only with replicate measurements.

Random errors are associated with the precision of the analytical method. According to the ICH definition, precision can be divided into:

- *Repeatability,* defined as the closeness of agreement between independent test results obtained with the same method on identical test material under the same conditions (same laboratory, same operator, same equipment within short intervals of time).
- *Reproducibility,* defined as the closeness of agreement between individual test results obtained with the same method on identical test material but in different laboratories with different operators using different equipment and not necessarily in short intervals of time.

- *Intermediate precision,* defined as the closeness of agreement between individual test results obtained with the same method on identical test material but in one laboratory, where operator and/or equipment and/or time are changed.

During the analysis of variance (ANOVA) the intrabatch and interbatch precision values are calculated as part of the validation process suggested by the 2000 Washington conference guidelines. Actually, while the definition of intrabatch precision corresponds to the ISO definition of repeatability, the interbatch precision calculated during the validation process is closer to the ISO definition of intermediate precision than to the definition of reproducibility. Indeed, in calculating the interbatch precision during method validation, it is usual to change only the time, sometimes the operator, rarely the instrument, hardly ever the laboratory. In general, because there are more steps involved in the assay process extended over several days, the intrabatch precision is less than the interbatch precision.

The separation of the two kinds of errors is useful in deciding which corrective action should be taken to improve the method performance during the development step [2,5]. Precision can be improved by increasing the number of replicates without changing the analysis procedure. However, if the bad performance of a method is due to bias, increasing the number of replicate measurements could not improve the method performance. In this case the method needs an investigation to understand the origin of bias and additional development.

To reduce the source of bias, it is important to have an authentic reliable reference standard, a properly maintained and calibrated balance, knowledge of the purity and form of the standard, calibrated glassware, calibrated pipettes, analyte-free matrix, and pooled and individual lots. Poor calibration, incorrectly or poorly prepared standards, interaction between analytes and container, and incomplete reactions will lead to bias. Some further precautions can be taken to minimize the possibility of errors, such as preparing QC samples in bulk (pools) and separately from the calibrators, which are prepared fresh everyday.

±15/20% Criteria. Both the 1990 and 2002 Washington conferences use the ±15/20% criteria for both accuracy and precision. Specific statistical calculation has been performed to understand the meaning and the probability of achieving the 15% criterion using different combinations of systematic and random errors [2]. The conclusion of this calculation has been that the ±15% criterion requires that the bias and the precision are ≤8% for five replicates. When the bias becomes larger, the precision must increase to meet the acceptance limits with sufficient probability, and vice versa. When both precision and bias are >8% ($n = 5$), the sample size must be increased (i.e., more than five replicates must be analyzed so that a reliable decision can be made to accept or reject an analytical method) [2]. Both the 1990 and 2002 Washington conferences require that the precision must be better than 15/20% for both intra- and interbatch. However, in practice the acceptance of a method when the intrabatch precision is close to 15/20% can lead the method to fail further validations of the performance. It has

125

been statistically calculated that it is unlikely in practice to obtain a reproducibility CV of 15/20% when the repeatability CV of the method is already 15/20% [2]. Since the differences between the repeatability CV and reproducibility CV can become quite large, it would be better to set separate limits for the repeatability and the reproducibility of the method.

Acceptance Criteria. Daily acceptance criteria are the performance standards against which applications of method are judged. Two approaches are used to define the acceptance criteria: (1) with respect to standard deviation of data obtained during validation, and (2) with reference to externally imposed requirements. These criteria should be established during method development or immediately following validation, prior to application of the method to study samples [1,3,5,11,14,17]. Aspects of the assay that could or should be included in acceptance criteria include:

- Accuracy
- Precision
- Number of standards that can fail an accuracy criterion
- Number of QCs that can fail an accuracy criterion
- Contamination of blank from carryover
- Maximum variation in the internal standard intensity
- Minimum acceptable internal standard intensity
- If curves bracket study samples
- How QCs should be distributed in the batch
- Extent of divergence
- Performance of system suitability samples

Internal Standard. The use of internal standard is critical in bioanalytical methods to improve precision and accuracy. The role of internal standard is to mimic the analyte of interest. It should be added before sample preparation/extraction to account for losses and errors introduced during the process. The more sample handling steps there are, the greater the error becomes, because errors are additive. In this case, the use of internal standard minimizes errors significantly.

The right internal standard should have similar chemical and physical properties to match those of the analyte of interest. In LC-MS and LC-MS/MS assays, the coelution of the internal standard represent the ideal situation. Therefore, both structural analogs and stable labeled isotope forms of the analyte can be used. Stable labeled isotope internal standards are an ideal choice of every mass spectrometric-based assay because they offer the highest precision. However, not all the stable labeled isotope forms of the analyte behave in the same way when used as an internal standard. Actually, the analog compounds may behave better than the wrongly labeled isotope internal standards.

Therefore, the "perfect" stable labeled isotope has to satisfy the following requirements:

1. It should incorporate ^{13}C, ^{15}N, and ^{18}O instead of deuterium (^2H) as the first choice to minimize the isotope effect and possibility of exchange. Multiple deuterium atoms can result in a significant isotope effect, to the point that the fully labeled form can be chromatographically resolved completely from the native. This means that the fully labeled form is not in the LC-MS or LC-MS/MS ion source at the same time as the analyte, and therefore this standard cannot control ion source events, limiting its ability to mimic the analyte behavior.

2. It should have more than two labeled atoms, to avoid isotopic contribution of the analyte.

3. It should have got high isotopic purity to avoid multiple peaks spaced by only 1 amu difference due to partially labeled forms.

4. The unlabeled isotope should be absent because it will affect the quantification of the analyte directly to produce a bias.

5. The labeled atoms must not undergo exchange at any point during the analysis.

The disadvantages of stable labeled isotope internal standard are its cost and availability. They are very expensive because they often require custom synthesis.

The amount of internal standard added should be similar to the amount of analyte contained in the sample. In fact, errors are minimized when the relative responses for analyte and internal standard are comparable. If a wide concentration range is to be measured, the amount of internal standard should be selected to maximize precision where it is critically important. As a general rule, with a six-level calibration curve, the internal standard concentration, should be between the second and third calibration standard concentration depending on the relative instrument response for analyte and internal standard and precision at the LLOQ level (i.e., internal standard should be closer to LLOQ if the precision of the LLOQ is low). Many bioanalytical laboratories use a written policy for rejecting sample with low internal standard even though it is not suggested by the 2000 Washington conference final report.

8.3.6 Dilution

Validation of dilution in biological matrix can be required during method validation or after the initial validation if some study sample concentrations are expected to be higher than the ULOQ and/or if the sample volume for some sample is less than required in the assay. Usually, dilution should be validated using at least five replicates in three separate analysis batches. The sample should be diluted with the same blank matrix type used in the validation and sample analysis, giving a resulting concentration in the upper half of the standard curve range. For limited quantity matrices, an appropriate proxy matrix may replace the original blank matrix. The possibility of using water for dilution has been investigated and is discussed in Section 8.3.10. Accuracy and precision for dilutions should be ≤15%.

8.3.7   Recovery, Carryover, and System Suitability Test

Recovery. In the 2000 Washington Conference final report it is clearly stated that knowledge of recovery is not essential to assay validation, but it does provide useful information about the real amount of analyte that is being analyzed. Assessment of recovery at every step of sample preparation and analysis in which losses may occur provides a powerful diagnostic tool to improve the method if needed. If a good internal standard has been chosen, the losses will have no impact on quantitation because they will be similar for analyte and internal standard and will thus annul each other. On the other hand, recovery is very important to verify if the internal standard really mimics and matches the analyte. The discovery of significant and inconsistent differences in recovery between analyte and internal standard at different steps of sample cleanup and analysis could indicate possible failure of the method during the validation.

For a better understanding of method performance, recovery should be divided in the following types and calculated independently:

- Total recovery
  - Comparison of either the peak heights or areas of extracted standard samples to peak height or areas obtained when the neat standard solution is injected at the same concentration as the extracted sample.
- *Extraction efficiency.* Recovery of the analyte from biological matrix after sample pretreatment (i.e., liquid–liquid extraction, solid-phase extraction, protein precipitation, etc.) to remove endogenous substances.
  - Comparison of the peak height or areas of extracted standard samples to the peak height or areas obtained when the extracted blank spiked with analyte is injected at the same concentration as the sample extracted.
- *Matrix effect* (enhancement or suppression). The effect of a biological matrix on the instrument response of the analyte [20].
  - Comparison of the peak height or areas of extracted blank spiked with analyte to peak height or areas obtained when the neat standard solution is injected at the same concentration as the extracted blank spiked sample.
- *Derivatization or reaction yield* (if applicable). It is used when the analyte is unstable or not sufficiently sensitive in the original chemical form for the techniques used.

The extracted blank spiked with analyte is an aliquot of blank matrix that is extracted and spiked with the appropriate neat standard solution prepared to be equivalent to an extracted standard.

If online extraction techniques are used, calculation of the different types of recovery could be particularly challenging. Different approaches has been used in literature to overcome this problem, such as using an external injection valve for adding the neat solution to the extracted blank from the online extraction column just on the head of the analytical column [9,10]. Although liquid matrix can be extracted directly, a solid matrix such as tissue needs to be disrupted in the presence of the solvent to allow complete dissolution. In this case the analyte degradation and incomplete extraction is possible.

Carryover. Carryover is due to a chromatographic peak at the retention time of the analyte or internal standard that has been detected in the following situation:

1. A blank or
2. In a STD_0 (extracted matrix sample that does not contain analyte but does contain internal standard at the level specified in the assay), or
3. Reconstitution solvent blank, which is caused by the residual analyte (or internal standard) from injection of the previous sample.

The evaluation of carryover in a batch is usually done by injecting a reconstitution solvent blank immediately after the ULOQ of the standard curve. A common acceptance criterion for carryover, in accordance with the 2000 Washington conference guidelines, is to reduce it to $\leq 20\%$ of the LLOQ. Carry-over should be evaluated during method development and its source investigated. It should be minimized by testing different rinse solvents and autosampler parameters, to reduce it to $\leq 20\%$ of the LLOQ using the peak areas (or $\leq 20\%$ of the internal standard concentration for the internal standard). If this condition is not met yet, the dynamic range of the standard curve should be modified, dropping the ULOQ so that $\leq 20\%$ carryover is easily achievable.

Even if a reconstitution solvent blank is normally used to evaluate carryover during validation and sample analysis, it is suggested to use all three types of blank as mentioned earlier for carryover evaluation during the method development. As considered previously (see Section 8.3.2) in the quantitative evaluation of carryover, linearity of detection together with sufficient accuracy and precision below LLOQ should be assumed. This assumption is questionable, and for this reason some bioanalytical laboratories have chosen different criteria to evaluate the carryover.

System Suitability. Although method validation is performed once at the end of method development, system suitability tests are performed on a specific system periodically (usually daily) or prior to each batch during validation and sample analysis to determine the system performance (see Chapter 13). During method development or/and upon completion of the validation, system suitability data should be evaluated and used to define acceptance criteria to use before starting sample analysis. System suitability tests include: (1) the reproducibility of retention time, (2) adequate sensitivity to quantify LLOQ (minimum detector response), (3) appropriate sensitivity to quantify ULOQ (within range of detector), and (4) chromatographic separation.

Usually, method-specific acceptance criteria are established for system suitability tests. For LC-MS/MS-based assay a possible general criteria for system suitability test is based on the use of at least five replicates of the neat solution concentration comparable with the closest calibration standard to LLOQ. To pass the criteria, this solution should give a precision for the five replicates $\leq 5\%$ (system fully equilibrated), and the sensitivity, as areas, should not be $\leq 40\%$ (still adequate sensitivity to quantify LLOQ) of the area of the first day of validation or sample analysis. If the sensitivity drops below 40% of the initial value, the LC-MS/MS should be retuned or recalibrated and the loss of sensitivity investigated.

8.3.8 Stability

The 2000 Washington Conference final report suggests testing stability comparing analytical results for stored samples with those for freshly prepared samples. However, other statistical approaches based on confidence limits for evaluation of analyte stability in a biological matrix can be used. Due to characteristic biological processes that could happen in an actual sample (in vivo) and not in a spiked sample (in vitro), additional stability investigation of samples from dosed subjects (in vivo) is suggested [3,4,11,15,19].

During method development it is important to start by investigating analyte stability. Analyte instability can arise for many reasons:

- Evaporation of volatiles
- Reaction with air (oxidation)
- Interaction with a container surface, such as:
  - Selective adsorption of analyte
  - Sticking to glass or plastic (e.g., proteins), especially when the analyte concentration is low
  - Leaching out contaminants: Glass may leach trace elements (sodium clustering MS analysis); plastic can produce plasticizers, which may confound MS analysis
- Photolytic decomposition
- Thermal decomposition
- Catalytic and enzymatic activity
- Physicochemical reactions with other sample components (protein binding)

This loss can be minimized or eliminated by the following steps:

- Lowering the temperature to inhibit degradative reactions, whether chemical or enzymatic:
  - Working on ice
  - Cooling the autosampler
  - Precooling vials, tubes, and any container
  - Keep the solutions and the sample at $-70°C$
- Protecting the analyte from light (amber bottle or aluminum foil)
- Adding a stabilizing agent (antioxidants or anticoagulant)

- Adjusting the pH to minimize analyte hydrolysis
- Adding enzyme inactivators or inhibitors
- Using chemical derivatization to stabilize the analyte
- Minimum sample degradation can be controlled by the addition of stable labeled isotope internal standard at time of collection [3,5,19].

Measurement of Metabolites. The elimination of drug usually happens through metabolites that are more polar than the drug itself to enhance their excretion via urine or bile. Some special situations exist in bioavailability/bioequivalence that require the determination of metabolites:

- When the parent drug cannot be measured in biological samples and only metabolite can be measured (e.g., when a prodrug is administered, it produces an active metabolite)
- When the parent drug along with active and/or inactive major metabolite(s) can be measured
- When more than one metabolite is present
- When the accumulation of metabolite is augmented (e.g., in the case of renal impairment)

All methods used for measuring metabolite(s) or parent drug and metabolite(s) level should be validated for that particular study matrix, with the same general parameters and criteria discussed above [3,23].

Stereoisomer Assays. There are many drugs that are administered as racemic mixtures. They may undergo stereoselective metabolism and/or elimination, and one isomer may be more active than the other. Therefore, there is the need to develop and validate bioanalytical assays for stereoselective determination in bioavailability/bioequivalence studies. All methods used for measurement of stereoisomer should be validated (with emphasis on stereospecificity). For bioequivalence studies of an existing racemic product, a stereospecific assay is not required if the rate and extent of profiles are superimposable (within the usual statistical boundaries) [3,23].

8.3.9 Documentation

A bioanalytical laboratory conducting pharmacology/toxicology and other preclinical studies for regulatory submissions should adhere to FDA's Good Laboratory Practices (GLPs) (21 CFR Part 58) (see also Chapter 13) [22]. Compliance with it is intended to assure the quality and integrity of the safety data filed. The GLP concept takes into account the fact that analytical measurements are the outcome of interactions among the sample (whose integrity must be assured), the instrumentation used to perform the weighing and measuring operations, and the analysts who perform these operations and subsequently process and interpret the raw data [6]. All of these interactions must be under well-defined control to fully satisfy GLP requirements. Both QA/QC monitoring of the scientific and technical procedures and full documentation of procedures and archiving of data are required in GLP regulations.

The bioanalytical laboratory should have a written set of standard operating procedures (SOPs) to ensure a complete system of quality control and assurance. The SOPs should cover all aspects of analysis from the time the sample is collected and reaches the laboratory until the results of the analysis are reported. All deviations from SOPs must be authorized by the study director and documented in the raw data. Significant changes in established SOPs must be properly authorized in writing by management.

The quality and integrity of the raw data should be strictly controlled and guaranteed. Raw data are any laboratory worksheets, records, memoranda, printed and handwritten notes, photographs, microfiche copies, computer printouts, magnetic media, dictated observations, recorded data from automated instruments, or exact copies that are the result of original observations and activities of a study. Overall, the term *raw data* covers all data necessary for reconstruction of the report of the study.

The documentation package for a method (method report) should be approved by management. The method report should include:

1. A complete list of equipment, conditions, reagents, and a description of reagent preparation required
2. A procedure for standard curve, validation, and quality control sample preparation
3. A procedure for sample extraction or sample treatment prior to analysis
4. A description of a system suitability test to be used prior to sample analysis, which will include acceptance/rejection criteria
5. The limits of the quantification range
6. Assay-specific acceptance and rejection criteria for a standard curve and QCs
7. A summary of the validation data, including the accuracy and precision results
8. A summary of the back-calculated standard curve values obtained during validation
9. Representative chromatograms of the plasma blank, plasma plus internal standard, system suitability solution, LLOQ, and ULOQ
10. A representative standard curve plot
11. Notebook references
12. Information supporting stability
13. Special method notes

8.3.10 Universal Bioanalytical Method Theory

The 1990 and 2000 Washington conference final reports and FDA guidance for the industry state that partial validation is required with a change in species within the matrix (see Section 8.2.9), and that calibration standards, QCs, and dilution QCs in the method validation should be in the same matrix of actual samples (see Sections 8.2.2 and 8.2.4). These requirements are based mainly on the fact that the analyte of interest, when in the same matrix (e.g., plasma) of different species (e.g., rat, mouse, dog, human, etc.), even if selectivity has been established, can still have different stability or different matrix effects. The *universal bioanalytical method* theory, based on the development and validation of a plasma matrix independent method [24,25], presents the following unique qualities:

- It can be used without changes for both animal species and human samples.
- It requires only a single validation for all plasma matrices.
- It allows the use of calibrators, QCs, and actual samples in different matrices in the same batch.
- It allows the use of dilution of biological matrices in water.

The possibility of replacing the plasma matrix diluent with water is particularly useful in preclinical studies where a very wide quantifying range is requested to accommodate the variable concentration of analyte in the study samples. The universal bioanalytical method employs:

- LC-MS/MS-based assays
- Linear fit for high-accuracy area ratio versus concentration
- Standard systematic method development template applicable to any compound
- Range template fixed to cover almost any ADME scientist's request
  - 0.5 to 500 ng/mL
  - ×10 dilution up to 5000 ng/mL
  - ×100 dilution up to 50,000 ng/mL
- Universal enzyme inhibitors for any type of plasma (same stability)

The principal advantages of this new approach are:

- Easy and standard procedure for method development, validation, and sample analysis:
  - Sample analysis is simpler and faster if samples of different matrices can be assayed together in a single batch using standards prepared in only one type of plasma matrix.
  - Simplification of the automated sample preparation and reduced possibility of clogging of the system probe tips due to plasma.
- High throughput (samples of different matrices in the same batch).
- Reduced cost per study:
  - Dilution in water lowers the expense of purchasing the blank matrix.
  - The least expensive matrix can be used for calibrators and QC preparation.

- Reduced and simplified paperwork:
  - Development and validation of only one analytical method is needed in this case to support all the studies in different species.

The feasibility of developing and validating plasma matrix–independent bioanalytical methods in the GLP environment should promote the production of new, upgraded SOPs concerning method validation and sample analysis, and it could be an input for the regulatory agencies for revisiting the present guidance in the bioanalytical field.

Methods. A plasma matrix–independent method has been obtained by adding specific enzyme inhibitors to all the various plasma matrices, and an insignificant matrix effect has been obtained using the following criteria:

- Calibration standards, QCs, and samples should be diluted in water 5× or 10× (depending on LLOQ response) before sample processing to make the different matrices uniform.
- APCI source should be used in LC-MS/MS. An APCI source has less matrix effect than an ESI source or no matrix effect.
- Stable labeled isotope internal standard should be used to compensate differences between matrices. Normal-phase chromatography (traditional or with water as a strong eluant) should be used instead of reversed-phase chromatography. It has been noticed that in normal-phase chromatography, the matrix effect is minimized. The k' value in chromatography should be increased from 2 to 4 to 8. An acceptable chromatographic separation is needed (retention time, 3 to 5 min) to separate the analyte from the endogenous interference. Pre- and postpeak materials should be diverted. This procedure reduces the memory effect in the ion source.

Results. The possibility of developing and validating "universal" bioanalytical methods has been demonstrated using both commercial drugs and new chemical entities. The method development and validation for Gemcitabine (a cytidine analog nucleoside used in the treatment of several tumor types) and its deaminated metabolite will be used as an example to demonstrate the applicability of the universal bioanalytical method theory [26,27]. The Gemcitabine method satisfies all six criteria above. Therefore this method has an insignificant matrix effect. A large excess of tetrahydrouridine (THU) has been added to actual samples and validation samples to prevent enzymatic conversion of Gemcitabine to its metabolite. In addition to the human plasma validation, the method was cross-validated in dog, rat, minipig, and mouse using human plasma calibrators.

The decision tree and infusion tuning experiments used in the systematic method development for choosing the mobile phase to achieve the best sensitivity in LC-MS/MS and reducing the matrix effect are shown in Figures 8.1 to 8.3. The run time was 3.5 min using 96.5 : 3.5 acetonitrile/1 mM ammonium acetate as mobile phase. Only the peaks of interest were collected; all the rest were

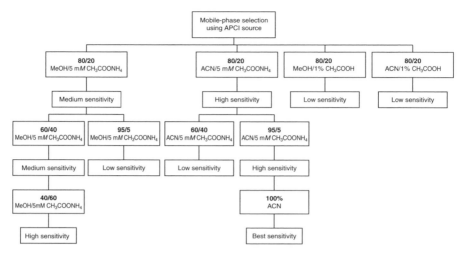

Figure 8.1. Systematic method development: mobile-phase choice decision tree.

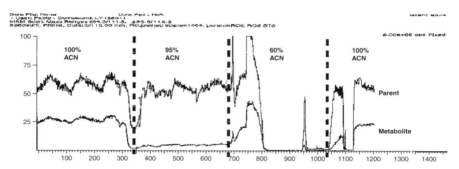

Figure 8.2. Systematic method development: mobile-phase choice using infusion experiment.

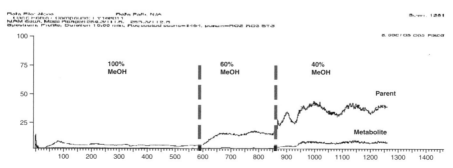

Figure 8.3. Systematic method development: mobile-phase choice using infusion experiment.

diverted to waste (Figure 8.4). The calibration range of assays using 100 μL of plasma was 0.5 to 50,000 ng/mL. The interbatch % RE for Gemcitabine was 2.67 for 10× dilution and 0.68 for 100× dilution in plasma and 2.97 for 10× dilution and 1.34 for 100× dilution in water (Table 8.1). The intrabatch % RE, maximum over all concentrations for the cross-validation using human calibrators was 12.72 (Human), 12.13 (Dog), 7.36 (Rat), 15.51 (Minipig), and 12.13 (Mouse)

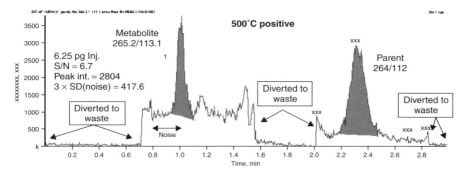

Figure 8.4. Representative mass chromatogram of an LLOQ level standard in human plasma for gemcitabine and its deaminated metabolite.

Table 8.1. Interbatch Accuracy and Precision for Gemcitabine and Its Deaminated Metabolite in Human Plasma Samples Diluted with Plasma versus Water

a. Parent

	4,027.680 10×-Dilution (Plasma)	40,276.800 100×-Dilution (Plasma)	4,027.680 10×-Dilution (Water)	40,276.800 100×-Dilution (Water)
Interday mean	4,135.183	40,549.703	4,147.459	40,815.612
Interday % RE	2.67	0.68	2.97	1.34
Interday % RSD	2.66	6.22	4.88	5.37
n	18	18	30	27

b. Metabolite

	4,009.280 10×-Dilution (Plasma)	40,092.800 100×-Dilution (Plasma)	4,009.280 10×-Dilution (Water)	40,092.800 100×-Dilution (Water)
Interday mean	4,232.519	41,568.796	4,210.142	41,549.593
Interday % RE	5.57	3.68	5.01	3.63
Interday % RSD	3.00	6.37	4.26	5.32
n	18	18	30	27

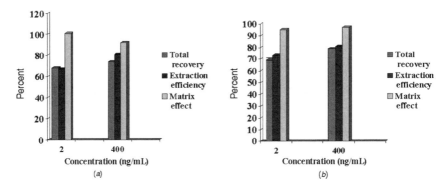

Figure 8.5. Total recovery, extraction efficiency, and matrix effect for (*a*) gemcitabine and (*b*) its deaminated metabolite from human plasma.

Table 8.2. Intrabatch Accuracy (% RE) and Precision (% RSD) for Gemcitabine and Its Deaminated Metabolite[a]

	Human	Dog	Rat	Minipig	Mouse
GB					
% RE	−12.72	12.13	7.36	15.51	12.13
% RSD	19.02	3.40	3.75	7.07	2.99
GBmet					
% RE	−16.57	11.98	14.57	13.77	12.57
% RSD	11.87	7.19	6.67	4.91	6.61

[a]Evaluated at 0.5, 250, and 500 ng/mL using calibrators in human plasma and QC standards in human, dog, rat, minipig, and mouse plasma. Human plasma: six replicates at each level quantified in four batches; animal plasma: six replicates at each level quantified in one batch for each species (dog, rat, minipig, mouse). Maximum over all concentration levels is reported.

(Figure 8.5). The matrix effect on the recovery was minimized to only 5 to 8.2% (Table 8.2).

Conclusions. An automated solid-phase extraction LC-MS/MS analytical method has been developed for quantifying Gemcitabine and its deaminated metabolite in human and animal plasma sample containing a large excess of THU. Since cross-validation of dog, rat, minipig, and mouse plasma samples using human plasma calibrators was successful for both parent drug and metabolite, the possibility of using human plasma calibrators to quantify analytes in animal matrices has been demonstrated together with the possibility of using water instead of matrix blank as diluent.

REFERENCES

1. P. V. Shah, K. K. Midha, S. Dighe, I. J. McGilveray, S. P. Skelly, A. Yacobi, T. Layloff, C. T. Viswanathan, C. E. Cook, R. D. McDowall, K. A. Pittman, and S. Spector, *Pharm. Res.*, **9**, 588–592, 1992.

2. C. Hartmann, D. L. Massart, and R. D. McDowall, *J. Pharm. Biomed. Anal.*, **12**, 1337–1343, 1994.

3. P. V. Shah, K. K. Midha, J. W. A. Findlay, H. M. Hill, J. D. Hulse, I. J. McGilveray, G. McKay, K. J. Miller, R. N. Patnaik, M. L. Powell, A. Tonelli, C. T. Viswanathan, and A. Yacobi, *Pharm. Res.*, **17**, 1551–1557, 2000.

4. FDA Guidance for Industry, *Bioanalytical Method Validation*, May 2001 (available from *www.fda.gov*).

5. A. Yergey, J. P. Gale, and M. W. Ducan, *Principles of Quantitative Mass Spectrometry*, ASMS Short Course, 2001.

6. L. Huber, *Good Laboratory Practice; A Primer for High Performance Liquid Chromatography, Capillary Electrophoresis, and UV–Visible Spectroscopy*, Hewlett-Packard Publ. 12-5091-6259E, 1993.

7. G. K. Szabo, H. K. Browne, A. Ajami, and E. G. Josephs, *J. Clin. Pharmacol.*, **34**, 242–249, 1994.

8. A. M. Almeida, M. M. Castel-Branco , and A. C. Falcao, *J. Chromatogr. B*, 2002 (in press).

9. B. T. Hoffman, J. R. Perkins, and J. Henion, *A Strategy for the Measurement of Recoveries During On-line SPE/LC/MS/MS* (contact information on *www.advion.com*).

10. F. Garofolo, H. Pang, M. McIntosh, E. Wong, M. Kennedy, and A. Marland, Turbulent flow chromatography coupled with normal phase liquid chromatography: a novel approach in using the cohesive standard single column mode (TFC-NPLC-MS/MS), *Proceedings of the 18th Montreux Symposium*, 2001.

11. A. R. Buick, M. V. Doig, S. C. Jeal, G. S. Land, and R. D. McDowall, *J. Pharm. Biomed. Anal.*, **8**, 629–637, 1990.

12. B. L. Ackermann, M. J. Berna, and A. T. Murphy, *Curr. Top. Med. Chem.*, **2**, 53–66 2002.

13. M. A. Brooks and R. E. Weifeld, *Drug Dev. Ind. Pharm.*, **11**, 1703–1728, 1985.

14. G. Ayers, D. Burnett, A. Griffiths, and A. Richens, *Clin. Pharmacokinet.*, **6**, 106–117, 1981.

15. S. T. Karnes, G. Shiu, and V. P. Shah, *Pharm. Res.*, **8**, 421–426, 1991.

16. T. C. Alexander, K. L. Norwood, S. P. Bode, T. R. Ierardi, J. R. Perkins, and J. Henion, *High-Throughput Sample Analysis in the GLP Environment* (contact information on *www.advion.com*).

17. A. C. Mehta, *J. Clin. Pharm. Ther.*, **14**, 465–473, 1989.

18. L. A. Pachla, D. S. Wright, and D. L. Reynolds, *J. Clin. Pharmacol.*, **26**, 332–335, 1986.

19. U. Timm, M. Wall, and D. Dell, *J. Pharm. Sci.*, **74**, 972–977, 1985.

20. B. K. Matuszewski, M. Constanzer, and C. M. Chavez-Eng, *Anal. Chem.*, **70**, 882–889. 1998.

21. J. Burrows, N. Shearsby, J. Chiu, K. Watson, and A. Witkowski, *A Pre-validation Strategy for Evaluating Bioanalytical Methods*, No. 3186 (available from BAS Analytics, West Lafayette, IN).

22. *Code of Federal Regulations*, Title 21, *Food and Drugs*, Office of the Federal Register, National Archives and Records Administration, Washington, DC, 1994, Part 58 (available from New Orders, P.O. Box 371954, Pittsburgh, PA 15250-7954).

23. H. Pang, *Impact of Biotransformation in Bioanalytical Assays: CVG LC-MS Discussion Group*, Mar. 21, 2002 (available from *www.cvg.ca*).

24. F. Garofolo, H. Pang, M. McIntosh, E. Wong, M. Kennedy, and A. Marland, Universal bioanalytical method theory: UBM, *Proceedings of the 47th ICASS*, 2001 (available from *http://orchard.uwaterloo.ca/conferences/icass2001/Main.htm*).

25. F. Garofolo, H. Pang, M. McIntosh, E. Wong, M. Kennedy, and A. Marland, A dream close in becoming true: "UNIVERSAL" bioanalytical method development and validation using LC-MS/MS, *Proceedings of the AAPS Annual Meeting*, 2002 (available from *www.aapspharmaceutica.com*).

26. F. Garofolo, H. Pang, M. McIntosh, E. Wong, M. Kennedy, and A. Marland, Quantitation of a cytidine analog and its deaminated metabolite in animal and human plasma by LC-MS/MS, *Proceedings of the 49th ASMS Conference on Mass Spectrometry*, 2001.

27. F. Garofolo, H. Pang, M. McIntosh, E. Wong, M. Kennedy, and A. Marland, "Aqueous dilutions of biological matrices: a simplified approach for extending the LC-MS/MS quantifying range, *Proceedings of the 49th ASMS Conference on Mass Spectrometry*, 2001.

9

PROCUREMENT, QUALIFICATION, AND CALIBRATION OF LABORATORY INSTRUMENTS: AN OVERVIEW

HERMAN LAM, PH.D.
GlaxoSmithKline Canada, Inc.

9.1 INTRODUCTION

The reliability of analytical data generated from chemical and physical analyses is critically dependent on three factors:

1. The validity of the analytical methods used
2. The reliability of the instruments used for the experiments
3. Proper training of the analysts

These three factors, linked together by cGMPs (current Good Manufacturing Practices), provide the fundamental assurance to the quality of the data. In this chapter, a systematic approach to ensure the reliability of the instruments used in the analyses is presented.

Nonreliable instruments can be a major source of error in all analyses. Analytical data generated from instruments that are not properly qualified or not calibrated with traceable standards are questionable and hence will be challenged. Regulatory agencies in most countries demand the use of calibrated instruments

Analytical Method Validation and Instrument Performance Verification, Edited by Chung Chow Chan, Herman Lam, Y. C. Lee, and Xue-Ming Zhang
ISBN 0-471-25953-5 Copyright © 2004 John Wiley & Sons, Inc.

for data generation. The International Organization for Standardization (ISO) and International Conference of Harmonization (ICH) have similar requirements.

1. FDA cGMP requirements [Code of Federal Regulations (CFR), Subpart I: Laboratory Controls, S211.160 (b)(4)]: [1]

The calibration of instruments, apparatus, gauges, and recording devices at suitable intervals in accordance with an established written program containing specific directions, schedules, limits for accuracy and precision, and provisions for remedial action in the event accuracy and/or precision limits are not met. Instruments, apparatus, gauges, and recording devices not meeting established specifications shall not be used.

2. International Organization for Standardization ISO/IEC 17025, General Requirement for the Competence of Testing and Calibration Laboratories, Section 5.5.2: [2]

Equipment and its software used for testing, calibration and sampling should be capable of achieving the accuracy required and shall comply with specifications relevant to the testing and/or calibrations concerned.

3. International Conference of Harmonization (ICH) Quality Guideline Q7A, Section 5.30: [3]

Control weighting, measuring, monitoring and test equipment that is critical for assuring the quality of intermediates or APIs (active pharmaceutical ingredients) should be calibrated according to written procedures and an established schedules.

The instrument qualification and calibration records are among the most frequently requested items in regulatory inspections. It is vital for the pharmaceutical, biotechnology, environmental, food, cosmetic, and chemical laboratories to maintain a rigorous instrument qualification and performance

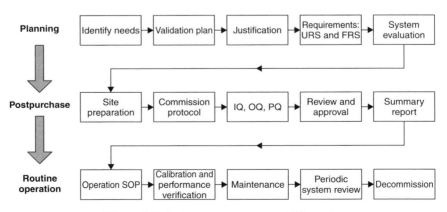

Figure 9.1. Instrument procurement life cycle.

verification program. Instrument qualification and calibration records are also among the most cited observations. It is very important to realize that the users (not the instrument providers) are ultimately responsible for the instruments in a laboratory.

A good program to support instrument procurement, validation, and operation starts from the planning to bring a new instrument into a laboratory to the decommissioning of the instrument at the end of its useful life. The program usually goes through three phases: (1) prepurchase planning phase, (2) postpurchase phase, and (3) routine operation phase. The activities associated with each phase are shown in Figure 9.1.

9.2 PREPURCHASE PHASE

9.2.1 Identification of Needs

The rationale for bringing a new instrument into the laboratories should be well founded. The benefits of acquiring the instrument, such as increasing productivity, meeting a specific need or regulatory compliance requirements, or enhancing the capability of the laboratories, should justify the expenditure of valuable resources required to bring the instrument in house and to support its operation.

9.2.2 Planning

Instruments to be used in a GMP environment have to be validated prior to regular use. It is useful to develop the instrument implementation plan as a plan to outline the steps necessary to validate the instrument and to support the use of the instrument through its functional life in the laboratories. The implementation plan should highlight the activities in the validation of the system and will be the guiding document of the project. Typically, the validation plan includes:

- The objectives and scope of the project
- A description of the system to be implemented
- The location of the instrument to be installed
- A validation approach
- The roles and responsibilities of the implementation team
- A project time line and milestones
- Constraints: physical (space, facility), financial, and project deadline
- References of relevant information

Note: The instruments used in the laboratories vary significantly in the design and operation complexity. The validation requirements should reflect the level of complexity of the instrument. It is obvious that the amount of effort required to validate a simple pH meter will be different from that required for a fully automated dissolution system. The Good Automated Manufacturing Practice (GAMP)

Table 9.1. GAMP Classification for Laboratory Instruments

GAMP Classification	Examples
A. Simple COTS instrument (commercial off-the-shelf)	Balance, pH meter, digital thermometer
B. Complex COTS instrument	HPLC firmware, GC firmware
C. Simple SCADA (supervisory control and data acquisition) system	HPLC, GC, FTIR, UV
D. Complex configurable SCADA system	Automated workstations for tablet assay and dissolution
E. Bespoke (custom) SCADA system	Custom-designed automation systems

Table 9.2. Validation Activities for Instruments in Each GAMP Class

Activity	Class				
	A	B	C	D	E
Planning	Yes	Yes	Yes	Yes	Yes
Requirements	Yes	Yes	Yes	Yes	Yes
Supplier audit	No	Evaluate	Evaluate	Yes	Yes
Evaluation	Yes	Yes	Yes	Yes	Yes
Design qualification	No	Evaluate	Evaluate	Yes	Yes
Qualification: IQ, OQ, PQ	Yes	Yes	Yes	Yes	Yes
Validation report	Yes	Yes	Yes	Yes	Yes

[4] developed by the International Society of Pharmaceutical Engineers (ISPE) provides a useful reference for classification for laboratory instruments according to their complexity (Table 9.1) and the related activities required to validate the instruments (Table 9.2).

9.2.3 Budgeting and Justification for Acquiring New Instruments

It is necessary to capture all the costs required in implementing an instrument in a facility to avoid serious budget shortfalls. The cost should include at least:

- The purchasing cost of the instrument and accessories.
- Applicable taxes associated with the purchase.
- *Site preparation.* It can be a significant percentage of the total cost, depending on the scope of the building modification that is required.
- *System qualification.* Qualification protocols and the service to execute the qualification protocol may not be included in the cost of the instrument and have to be purchased separately from the instrument vendor.

- *Training.* There may be a need to send users to the instrument supplier to take operation and maintenance training. The traveling and tuition should be included in the budget.
- *Contingency funding.* This is included to cover any unforeseeable expenses. which is typically 10% of the sum of the cost of the foregoing items. Building modification changes during site preparation can lead to significant cost overruns.

Different emphasis may be used when justifying a new piece of instrument for use in a quality control laboratory as compared to an R&D laboratory. In a quality control environment, the justification is based primarily on the need and also on the return on investment. For R&D laboratories, the capability enhancement potential of the new instrument is also a major consideration. The investment in new R&D instruments may not have an immediate return.

9.2.4   Requirements

User Requirements. The users first need to decide on the major tasks the instrument should be able to accomplish. For example, if users need an HPLC for routine product release testing, an HPLC system with a variable-wavelength UV detector, isocratic pump, and an autoinjector is likely to be sufficient for routine assay of main active ingredient(s) in a pharmaceutical dosage form. However, if the HPLC system is intended to be used for stability-indicating assay or impurities assay, a system with a gradient pump that provides a wider choice of solvent strength for better separation and a more sensitive detector may be required. These high-level requirements, such as a HPLC system for routine product testing or stability studies, are captured in the user requirements. The ability of the system to perform tasks outlined in the user requirements has to be demonstrated in the performance qualification phase of the qualification process.

Caution should be used when putting together the user requirements since it will have a major impact on the amount of work required for the system qualifications. The more tasks that are specified in the user requirements, the more work has to be done during the qualification process to demonstrate that the instrument is capable of fulfilling the requirements. If the system is capable of performing additional tasks that are not required by the user requirements, there is no need to validate those tasks.

Functional Requirements. Based on the user requirements, more detailed functional requirements can then be defined. Take as an example a gradient HPLC system with UV–Vis detection required to run a stability-indicating method. The functions of each of the hardware and software components required to perform the tasks in the user requirements should be specified. The functional specifications typically include:

Hardware Components

- *Functions of each component and operation range:*

- *Pump:* flow rate range, gradient mixing mechanism, gradient accuracy, and solvent-delivering capability
- *Detector:* wavelength range, wavelength switching, sensitivity, and resolution (slid width)
- *Autoinjector:* injection volume range, sample capacity, level of carryover, precision, and temperature range if refrigeration of samples is required
- *Column temperature control:* temperature range
- *Operation environment of the system:* building temperature and humidity range
- *Site requirements to support the operation of the instrument:* power supply (voltage and current) and ventilation
- *Health and safety requirements:* electrical and mechanical safety
- *Uninterrupted power supply (UPS):* the length of the support during power failure and the power rating of the UPS

Software

- Computer operation system and network requirements if required
- *User interface:* operation modes and setup
- *System interface with hardware:* supervisory control and data acquisition (SCADA) program
- Data type and memory capacity
- *21 CFR Part 11 Electronic Records and Electronic Signatures (ERES):* system security, data integrity and tracibilty, audit trail, and archive
- System recovery

Design Qualification. For a commercial system, users generally have very little or no input into the design of the instrument. The design qualification in this case outlines the user and functional requirements and the selection rationale of a particular supplier. For a custom-designed system, the design qualification outlines the key features of the system designed to address the user and functional requirements.

9.2.5 System Evaluation

In addition to the factors already discussed, there are several others that need to be considered:

- It is obvious that the instrument has to be able to perform the tasks detailed in the user requirements.
- *Cost.* It is best to strike a balance between the cost and the performance of an instrument. The least expensive instrument may not be the best investment; the most expensive instrument may not be the best instrument for a particular operation.

- *Ease of use.* Simplicity is beauty. Purchasers should think about the general background of potential users. Not all users are ready to tackle very complicated operations, due to time constraints and training.
- *Vendor's reliability.* The vendor that supplies the instrument should have a track record of providing high-quality instruments and after-sale support. A vendor audit should be conducted for a new instrument supplier to evaluate the company's ability to build high-quality products. Purchasing an instrument from a financially unstable vendor is risky.

Most modern instruments are controlled by computer. It is important to have assurance from the vendor that the software and hardware were developed according to industrial standards such as the IEEE software development guide or the Good Automated Manufacturing Practice (GAMP) guide developed by the International Society of Pharmaceutical Engineers (ISPE). The software development life-cycle approach should have been used during development. Key quality assurance procedures, such as change control, security, management involvement, and off-site storage should be in place [5–10]. Computer system validation is discussed in Chapter 17.

9.3   POSTPURCHASE PHASE

9.3.1   Site Preparation

Users should study the site preparation guide from the vendor. Careful planning is required to ensure that the necessary preparations to house the new instrument in the laboratory have been completed prior to installation. Insufficient site preparation causes major inconvenience and long delays in the installation process. It is a waste of time and money to have the service engineer show up in the laboratory but not be able to do anything, due to poor site preparation. It is a common mistake to underestimate the effort and time required for site preparation.

Following are the key considerations for site preparation:

- *Physical dimensions.* Make sure that there is enough space to accommodate the instrument and accessories and that the bench is strong enough to support the instrument. The door should be wide enough for a large instrument to pass through.
- *Suitable operational environment.* Proper temperature, humidity, and vibration control must be maintained for the instrument.
- *Utilities.* Some instruments require one or more of the following utilities: high-voltage or high-current supply, special electrical plug to handle the high voltage and current, gas (helium, nitrogen, and air) supply, computer network connection, special ventilation and enclosure, water supply and drainage. *Example:* The installation of an inductively coupled plasma spectrometer requires a high electrical current supply and a special electrical plug. Installation of LC-MS systems may require installation of a

nitrogen generator. Engineering work to put in these necessary utilities takes some time.

- *Health and safety requirements.* Special licenses are required to operate an instrument that uses radioactive substances (e.g., the radioactive sources for the electron capture detector)

  In some countries, even the radioactive device used to reduce static charge for balances may require a license to operate. The x-ray sources for a crystallography system also merit close scrutiny.

- *Electrical safety.* Most common commercial instruments have been certified for electrical safety by an organization such as UL (Underwriters' Laboratory), CE, or CSA (Canadian Standards Association). However, certain less common instruments or custom-built instruments may have to go through a certification process at the time of installation. Certification required for electrical safety may take time to complete. A protective shield or casing may be required for automated systems with robotic arms for sample manipulation, to protect operators.

9.3.2   Qualifications

Instrument qualification is required to establish the functional capability and reliability of a system for its intended use in a suitable environment. Instrument qualification can be divided into three stages: installation, operation, and performance qualifications. A qualification protocol that provides details about the system, the scope and constraints of the qualification, qualification tests, test procedures, and acceptance criteria should be available for review and approval before qualification begins. Sufficient time should be provided for review and approval. The protocol should also contain an exception log to record any out-of-specification results, investigation, and problem resolution.

Installation Qualification. Installation qualification (IQ) is a process to establish that the instrument was received as specified and installed properly according to the design requirements in an environment suitable for its operation. Proper installation is the first step to ensure proper functioning of equipment. Typical IQ activities include:

- Verify the hardware and software delivered against the shipping list.
- Inspect for visible damage.
- Verify the software and firmware versions.
- Document the model, configuration, and serial numbers of the system components.
- Document the model, configuration, and serial numbers of the computer system.
- Download all files in the application/control software and verify.
- Assure proper power-up or startup of the system components.

- Establish proper communication between the system components and computer control.
- Set up an instrument logbook.
- Calibrate the system modules if necessary.

Operation Qualification. Operation qualification (OQ) is the process of establishing that the instrument or system modules operate according to the functional requirements in a suitable environment. For an HPLC system, operation of the pump, injector, and detector will be tested at this stage. Typical OQ tests for HPLC modules and a UV–Vis spectrophotometer are as follows:

HPLC

- *Pump:* flow rate accuracy and gradient accuracy
- *Detector:* linearity of response, noise, drift, and wavelength accuracy
- *Injector:* precision, linearity, and carryover
- *Column heater:* temperature accuracy

UV-Vis Spectrophotometers

- Wavelength accuracy and reproducibility, stray light, resolution, photometric accuracy and reproducibility, noise, baseline flatness, stability, and linearity

In additional to testing the system components, a functional challenge that tests the system software operation and security should be undertaken. A predetermined set of instructions can be input into the system step by step. The system responses are then compared to the expected outcomes of the instruction to determine any problems with its execution. Some vendors will provide a standard set of data that can be processed by the system to verify its data-handling capability.

Performance Qualification. Performance qualification (PQ) is a process to demonstrate that an instrument can fulfill the requirements outlined in the user requirements. The PQ can be demonstrated by running a typical application in the user requirements which requires the system components to function together properly to deliver the expected test results. The flow of the sequence of events from requirements preparations to system delivery (or system building) to instrument qualification, and the relationship between the requirements and the qualification processes, are outlined in Figure 9.2.

9.3.3 Data Review and Summary Report

There may be times when some test results fail to meet acceptance criteria. The nature and cause of the failure must be investigated. Based on the results of the investigation, the impact of the failures can be assessed. The cause of the failure can be instrument related or due to operator error. Once the cause of the failure has been identified, action can be taken to rectify the problem. The test should

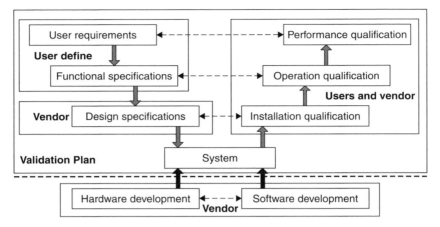

Figure 9.2. Relationship between requirements and qualification processes.

be repeated to demonstrate that the system's operations can fulfill the acceptance criteria. The failure, the investigation, the actions to rectify the failure, and the retest must be documented in the exception log.

After the qualification testing, a summary report should be written to summarize validation activities with a conclusion as to whether the instrument is suitable to be put into routine use. The report should highlight the objective and scope of the validation project, all the qualification test results, test exceptions, and a recommendation for system acceptance. All the test procedures, test results, and the summary report must be reviewed and approved by quality assurance and/or management before product use.

The entire qualification process has to be documented. *If it is not documented, it is just a rumor!* The records should be stored in a secure place because the records are the only proof that the qualification has been carried out. Systematic filing allows fast and easy access of records during an audit. Rapid record retrieval helps to convey a positive impression to auditors that the system is under control. It would a pity if all the hard work had been done but the records could not be presented quickly and efficiently during an inspection.

Users should get involved in the qualification process whenever possible. The qualification exercise provides a very good learning opportunity for users to work with the service engineer and learn more about the system's design and functions. Users will gain valuable experience with regard to operation and maintenance of the instrument.

9.4   ROUTINE OPERATION PHASE

9.4.1   SOP

After the instrument is qualified, it can be used to generate analytical data. A standard operating procedure (SOP) has to be written for the new instrument.

The operation instruction, maintenance, and calibration should be included in the SOP. It is not necessary to copy the entire operation manually into the SOP. Writing simple instructions with references to the related sections in the manual is a better way. The frequency and tasks to be performed during maintenance should be stated in the maintenance section. The tests required to calibrate the instrument, the acceptance criteria, and the frequency for each test should be included in the calibration section of the SOP. Definitions of major and minor repairs which necessitate partial or full system requalification should be included as well. For example, the replacement of a UV lamp in a UV detector does not require full requalification whereas replacement of a circuitry board will.

Good system maintenance starts with the users. Proper care, which can be as simple as a good system rinsing and cleanup after use, can reduce unwanted system failure in the middle of a run and extend the useful life of the instrument. Preventive maintenance is a good investment that will save valuable time and money in the long run.

9.4.2 Calibration and Maintenance

The cGMP requirements dictate that the calibration of instruments should be performed at suitable intervals in accordance with an established written program. Instruments not meeting established specifications must not be used. Each instrument should have a calibration sticker with information related to the status of the system, when the calibration was performed, who did the calibration, and the next calibration date. A systematic program is required to maintain the instruments in a state of calibration. The following points should be considered when setting up an instrument calibration and maintenance program.

- Responsibilities of the personnel involved in the calibration of the equipment
- Frequency of calibration for each type of instrument if it is not covered in the operation SOP of the instrument
- Review and approval of calibration data
- Procedure to issue calibration stickers (database application can be used to track the status of the instrument and help coordinate the calibration date)
- Documentation requirements of the calibration and record keeping
- Central filing for instrument-related records
- Remedial actions in the event of calibration failure
- Procedure to notify users and obtain impact assessment in case of calibration failure

It is very important to maintain good use and service records for the instrument for cGMP purposes. CFR 211.194 [8] specifically requires complete records to be maintained of the periodic calibration of laboratory instruments, apparatus, gauges, and recording devices. The records of use allow users to be notified in case of system or calibration failure. The user may have to do an impact assessment to determine whether the failure would have affected the reliability of the results generated by the system. The service records will also provide useful information about the system, which may simplify the troubleshooting effort in some instances.

The terms *calibration* and *performance verification* are very often used interchangeably. Calibration involves measuring and adjusting the instrument response using known standards. Performance verification verifies the operation and performance characteristics of an instrument against a predetermined set of requirements. Calibration is a part of performance verification.

There is a common misconception that running system suitability before the analysis can replace the need for regular instrument calibration. System suitability demonstrates only that the instrument is suitable for a particular application at the time of analysis. It cannot reveal marginal system performance. For example, the system suitability test for an HPLC assay using UV detection is not likely to pick up a wavelength accuracy problem since both the standards and the samples are quantitated at the same wavelength. System suitability testing is method specific, whereas system calibration verifies the general performance of the instrument.

9.4.3 Periodic Review

The performance of an instrument should be reviewed on a regular basis to ensure that the instrument is reliable and continues to comply with the requirements specified in the user requirements. The review should determine whether the instrument is maintained in a validated state. The review should include records for use, maintenance, services, and performance verification testing. In case the records indicate that the instrument is more prone to certain types of failure, preventive maintenance may be desirable to avoid system failure during operation. The review can also provide useful information to prioritize instrument replacement.

9.4.4 Change Control

Changes to the hardware, software, and firmware should be evaluated for the potential impact that the changes may introduce. The overall planing, evaluation, justification, implementation, testing, and approval processes are generally referred to as *change control*. Change control provides supporting evidence that the results expected from the changes have been achieved, and that the system complies with the specifications and remains in a validated state. The change control plan should be approved by laboratory management and quality assurance personnel prior to its implementation. Upon the completion of the change and related testing, the entire process has to be reviewed and approved prior to releasing the system for routine use.

9.4.5 System Decommissioning

When an instrument is no longer required in the laboratory and has to be decommissioned, all related records, such as the instrument binder or logbook, manuals, and operation software should be archived. Record in the equipment binder or logbook the reason for the decommissioning and the effective date of removal.

9.5   SUMMARY

A systematic approach to managing the entire life cycle of laboratory instrumentation, from procurement to decommissioning, is a good business practice for smooth laboratory operation. Doing things right the first time will help save time, money, and resources and avoid preventable instrument failures. A systematic approach conveys confidence to auditors during laboratory inspections. Most important, a systematic approach to make sure an instrument is functioning properly and adequately according to the intended requirements is one of the most important factors in ensuring the quality and reliability of data generated by the instrument.

REFERENCES

1. *Current Good Manufacturing Practices for Finished Pharmaceuticals*, 21CFR, Parts 210 and 211, 1995.
2. ISO/IEC 17025, *General Requirements for the Compliance of Testing and Calibration Laboratories*, 1999.
3. *International Conference on Harmonization (ICH) Quality Guideline Q7A*, 2000.
4. *GAMP (Good Automated Manufacturing Practice) Guide for Validation of Automated Systems in Pharmaceutical Manufacturing*, Version 4, 2002.
5. L. Huber, *Validation and Qualification in Analytical Laboratories*, Interpharm Press, Englewood, Co, 1999.
6. L. Huber, Quality assurance and instrumentation, *Accredit. Qual. Assur.*, **1**, 24, 1996.
7. L. Huber, Equipment qualification in practice, *LC-GC*, Feb. 1998.
8. M. Freeman et al. Position paper on the qualification of analytical equipment, *Pharm. Technol. Eur.*, **40**, 1995.
9. L. Huber, *Validation of Computerized Analytical Systems*, Interpharm Press, Englewood, Co, 1995.
10. W. Furman, T. Layloff, and R. Tetzlaff, Validation of computerized liquid chromatographic system, *J. AOAC*, **77**, 1314, 1994.

10

PERFORMANCE VERIFICATION OF UV–VIS SPECTROPHOTOMETERS

HERMAN LAM, PH.D.
GlaxoSmithKline Canada, Inc.

10.1 INTRODUCTION

UV–Visible (UV–Vis) spectrophotometry is commonly used in analytical laboratories for qualitative and quantitative analyses. The popularity of the UV–Vis technique stems from its ease of use and the speed of the analysis. The concentration of a sample can be calculated from the UV–Vis absorbance data using Beer's law. UV–Vis experiments are usually very simple to execute, but analysts sometimes take for granted the importance of the proper spectrophotometer performance qualifications required to generate reliable data. The performance requirements of spectrophotometers vary according to the nature of the tests and the design of the instrument. Certain performance characteristics will affect scanning instruments more than diode array instruments, and vice versa. Scanning instruments with a double-beam design generally provide better resolution and stability. Diode array instruments in general have the advantages of speed and better ruggedness in wavelength accuracy. Performance characteristics that affect the reliability of UV–Vis measurement and the verification tests are discussed in this chapter. These discussions focus on the instrument performance characteristics that are required to meet the regulatory requirements for pharmaceutical analysis.

The performance verification tests required by major pharmacopoeias for UV–Vis spectrophotometers are listed in Table 10.1 [1–4]. The required performance tests include wavelength accuracy, stray light, resolution, and photometric

Analytical Method Validation and Instrument Performance Verification, Edited by Chung Chow Chan, Herman Lam, Y. C. Lee, and Xue-Ming Zhang
ISBN 0-471-25953-5 Copyright © 2004 John Wiley & Sons, Inc.

Table 10.1. Performance Tests for UV–Vis Spectrophotometers

Performance Tests	Standards	Pharmacopoeia[a]			
		USP 26	BP 2001	EP 4 Ed.	JP (XIV)
Wavelength accuracy	Deuterium lamp	√	√	√	√
	Mercury vapor lamp	√	√	√	√
	Holmium oxide glass filter	√			√
	Didymium glass filter	√			√
	Holmium oxide in $HClO_4$	√	√	√	
Wavelength reproducibility					√
Stray light	Potassium chloride solution (200 nm)		√	√	
	Potassium iodide solution (220 nm)	√			
Resolution	Toluene in hexane solution		√	√	
Photometric accuracy	Neutral-density glass filters	√			
	Metal on quartz filters	√			
	Potassium dichromate solution	√	√	√	√
Photometric reproducibility	Potassium dichromate solution				√
Noise					
Baseline flatness					
Stability					
Linearity					

[a]USP, United States Pharmacopoeia; BP, British Pharmacopoeia; EP, European Pharmacopoeia; JP, Japanese Pharmacopoeia.

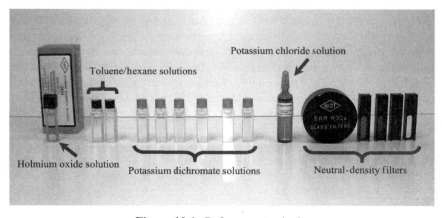

Figure 10.1. Reference standards.

accuracy [5,6]. There are other performance characteristics, such as noise, baseline flatness, and stability, that will affect the performance of spectrophotometers as well. Some reference standards commonly used in the performance verification of UV–Vis spectrophotometers are shown in Figure 10.1.

10.2 PERFORMANCE ATTRIBUTES

10.2.1 Wavelength Accuracy

Wavelength accuracy is defined as the deviation of the wavelength reading at an absorption band or emission band from the known wavelength of the band. The wavelength deviation can cause significant errors in the qualitative and quantitative results of the UV–Vis measurement. It is quite obvious that if the spectrophotometer is not able to maintain an accurate wavelength scale, the UV absorption profile of the sample measured by the instrument will be inaccurate. The true λ_{max} and λ_{min} of the analyte cannot be characterized accurately.

In addition to the qualitative problem, wavelength deviation affects the quantitative measurements in terms of accuracy and sensitivity. Typically, most UV–Vis assays will specify that the measurements are to be taken at λ_{max}. For an analyte with a fairly broad absorption profile, a slight shift in wavelength will have very little effect on the absorbance value. However, there are situations which necessitate that measurements be taken at the upslope or the downslope of the absorption profile to avoid interference, to remove the need for sample dilution, or to increase the dynamic range of the measurement. A small deviation in wavelength, especially for analytes with a steep slope in the absorption profile, will cause a large change in the absorbance value of the measurement, and the accuracy of the quantitation will be affected (Figure 10.2). Significant wavelength deviation will also steer measurement away from the optimal wavelength at λ_{max}, which effectively reduces the extinction coefficient and hence the sensitivity of

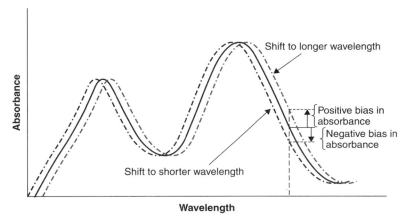

Figure 10.2. Effect of wavelength accuracy on UV–Vis measurements.

the measurement. Diode array instruments with no moving parts in the spectrograph are usually less susceptible to wavelength deviation than are scanning instruments.

Tests. Wavelength accuracy verification is checked by measuring a known wavelength reference standard with well-characterized absorption or emission peaks and comparing the recorded wavelength of the peak(s) against the value(s) listed in the certificate of that reference standard. There are many standards that are commonly used to verify the wavelength accuracy of a spectrophotometer. Spectra of some commonly used wavelength standards such as a deuterium lamp, mercury vapor lamp, holmium oxide filter, and holmium oxide solution (4% holmium oxide in 10% perchloric acid in a 1-cm cell) are illustrated in Figure 10.3a–d, respectively. The advantages and disadvantages of some commonly used wavelength verification standards are summarized in Table 10.2.

Holmium oxide solution is commercially available in a sealed 1-cm cuvette. It is a very convenient and versatile wavelength standard. The standard is suitable for UV–Vis spectrophotometers typically used in pharmaceutical laboratories, with spectral bandwidth ranging from 2 to 0.5 nm. The certified wavelength values of the peaks are listed in Table 10.3. When a spectral bandwidth of 0.5 nm or narrower is used in a measurement, the position of peak number 10 at about 451 nm becomes undefined, due to peak splitting as a result of enhanced resolution.

Acceptance. ±1 nm in the UV range (200 to 380 nm) and ±3 nm in the visible range (380 to 800 nm). Three repeated scans of the same peak should be within ±0.5 nm.

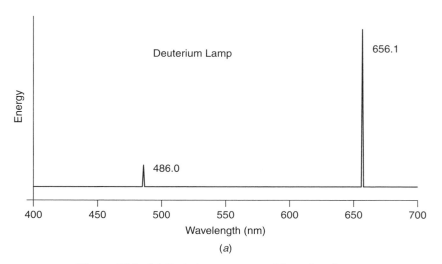

Figure 10.3. (*a*) Emission spectrum of deuterium lamp.

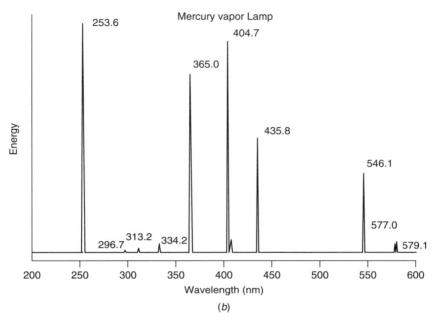

Figure 10.3. (*b*) Emission spectrum of mercury vapor lamp.

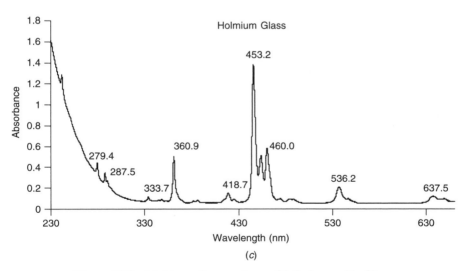

Figure 10.3. (*c*) Absorption spectrum of holmium oxide filter.

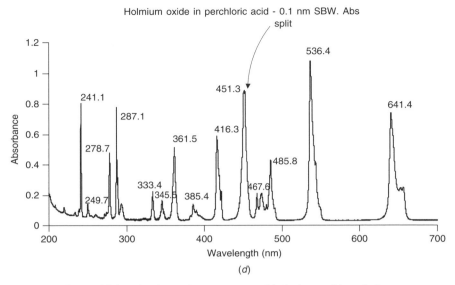

Figure 10.3. (*d*) Absorption spectrum of holmium oxide solution.

Table 10.2. Wavelength Standard Comparison

Wavelength Standard	Advantages	Disadvantages
Emission lines from deuterium lamp	Sharp spectral lines Present in the UV light source	Limited to visible wavelengths (486.0 and 656.1 nm)
Emission lines from mercury vapor lamp	Sharp spectral lines Covers both UV and visible regions	Not commonly built into the instrument
Holmium oxide and didymium filters	Easy to use NIST SRM available	Glass filter absorbs strongly below 300 nm: limited to mainly visible wavelengths (280–640 nm) Broad spectral lines Recertification required every two years
4% Holmium oxide in 10% perchloric acid	Easy to use NIST SRM available (2034) Covers both UV and visible regions (240–650 nm) Recertification required every eight years	Dependency on the resolution of the instrument Corrosive solution

Table 10.3. Wavelength Assignment: 4% Holmium Oxide in 10% Perchloric Acid

SRM 2034 Band	Spectral Bandwidth		
	0.5 nm	1.0 nm	2.0 nm
1	241.01	241.13	241.08
2	249.76	249.87	249.98
3	278.13	278.10	278.03
4	287.01	287.18	287.47
5	333.43	333.44	333.40
6	345.52	345.47	345.49
7	361.33	361.31	361.16
8	385.50	385.66	385.86
9	416.09	416.28	416.62
10	–	451.30	451.30
11	467.80	467.83	467.94
12	485.27	485.29	485.33
13	536.54	536.64	536.97
14	640.49	640.52	640.84

10.2.2 Stray Light

Stray light is defined as the detected light of any wavelength that is outside the bandwidth of the wavelength selected. The causes for stray light are scattering, higher-order diffraction of the monochromator, or poor instrument design [7,8]. Stray light causes a decrease in absorbance and reduces the linear range of the instrument. High-absorbance measurements are affected more severely by stray light. For example, consider the following measurement scenarios at 1% transmittance level:

$$\text{absorbance}(A) = -\log \text{transmittance } (T) \tag{10.1}$$

where T is the ratio of intensity of the transmitted light (I) and incident light (I_0). In the presence of stray light,

$$T = \frac{I + I_s}{I_0 + I_s} \tag{10.2}$$

where I_s is intensity of the stray light. If there is no stray light, the absorbance value as given by equation (10.1) should be 2.

The changes in absorbance due to different levels of stray light at 1% transmittance are estimated in Table 10.4. As the results indicate, stray light can cause a significant reduction in absorbance. At 1% transmittance, stray light at 1% of the incident light intensity can cause a 15% drop in absorbance. The effect of various levels of stray light on the absorbance value is shown graphically in Figure 10.4.

Table 10.4. Stray Light Estimation at 1% Transmittance

I_s (Stray Light)	I_i (Transmitted)	$I = I_s + I_i$	I_0	$T = I/I_0$	$A = -\log T$
0%	1%	1%	100.0%	0.0100	2.000
0.1%	1%	1.01%	100.1%	0.0101	1.996
0.5%	1%	1.05%	100.5%	0.0149	1.826
1%	1%	2%	101.0%	0.0198	1.703

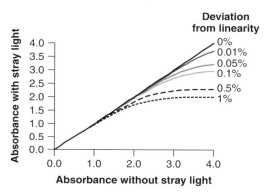

Figure 10.4. Effect of stray light on absorbance measurement.

The stray light problem causes a deviation from linearity at high absorbance. In general, the linearity of the absorbance response is limited by stray light at high absorbance and by noise at the low absorbance range. Absorbance values ranging from 0.3 to 1 AU are less susceptible to stray light and noise problems and therefore become the preferred absorbance range for UV–Vis analyses.

Tests. For the stray light test, various cutoff filters or solutions can be used to estimate the stray light contribution, depending on the wavelengths of interest. These cutoff filters or solutions have the optical characteristics to allow light at slightly higher wavelengths than the measuring wavelength to pass through and block off light at the measuring wavelength. Table 10.5 provides information on three test solutions for stray light measurement at 200, 220, and 340 nm. Scan the stray light testing solution in a 1-cm cell using air as the reference. Measure the transmittance or absorbance of the solution at the wavelength specified in Table 10.5. The spectra of aqueous KCl, NaI, and $NaNO_2$ are shown in Figure 10.5.

Acceptance. The transmittance of the solution in a 1-cm cell should be less than 0.01, or the absorbance value should be greater than 2.

Table 10.5. Stray Light Measurement

Spectral Range (nm)	Solution	Measurement Wavelength (nm)
175–200	Aq. KCl (12 g/L)	200
210–260	Aq. NaI (10 g/L)	220
300–385	Aq. NaNO$_2$ (50 g/L)	340

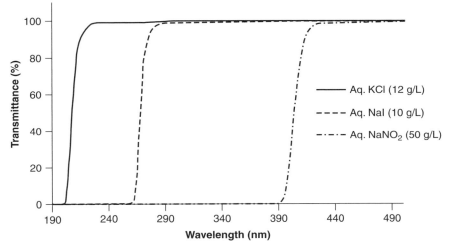

Figure 10.5. Stray light measurements.

10.2.3 Resolution

Insufficient resolution leads to a decrease in the extinction coefficient across the wavelength axis, and therefore inaccurate quantitation results. The sensitivity of the measurement is also compromised. From a qualitative point of view, the fine features in the spectrum may be lost. The resolution of a UV–Vis spectropho-tometer is related to its spectral bandwidth (SBW). The smaller the spectral bandwidth, the finer the resolution. The SBW depends on the slit width and the dispersive power of the monochrometer. Typically, only spectrophotometers designed for high-resolution work have a variable slit width. Spectrophotometers for routine analysis usually have a fixed slit width. For diode array instruments, the resolution also depends on the number of diodes in the array.

The resolution of the absorbance measurement depends on the ratio of the spectral bandwidth (SBW) of the spectrophotometer to the natural bandwidth (NBW) of the spectral band to be measured. The NBW is a physical characteristic of the analyte of interest. The accuracy is not likely to be affected if the ratio of NBW/SBW is greater that 20. If the ratio is less than 10 (SBW increasing), the

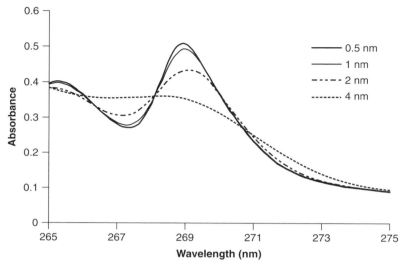

Figure 10.6. Spectral resolution at different slit widths.

measured spectrum becomes distorted. For example, the spectral changes from scans of a 0.02% v/v toluene-in-hexane solution using slit width of 0.5, 1, 2, and 4 nm are illustrated in Figure 10.6. At a 4-nm slit width, the peaks are very broad, and most of the spectral features are lost. In addition to the losses in qualitative information, the broadening of the peak also led to a reduction in the extinction coefficient of the peak, which in turn reduces the sensitivity of the quantitation.

Tests. A solution mixture of 0.02% v/v toluene in hexane (UV grade) is used to test the resolution power of the spectrophotometer. Hexane as the reference is scanned and then the spectrum of the resolution solution from 250 to 300 nm is obtained. The absorbance values of the λ_{max} at 269 nm and the λ_{min} at 266 nm are recorded (Figure 10.7).

Acceptance. The ratio of the absorbance at λ_{max} (269 nm) and absorbance at λ_{min} (266 nm) should be greater than 1.5.

10.2.4 Noise

Noise in the UV–Vis measurement originates primarily from the light source and electronic components. Noise in the measurement affects the accuracy at both ends of the absorbance scale. Photon noise from the light source affects the accuracy of the measurements at low absorbance. Electronic noise from the electronic components affects the accuracy of the measurements at high absorbance [8]. A high noise level affects the precision of the measurements and reduces the limit of detection, thereby rendering the instrument less sensitive.

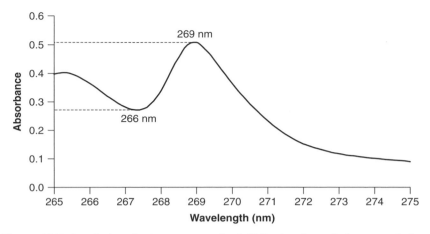

Figure 10.7. Resolution check: spectrum of a 0.02% v/v toluene in hexane solution.

Tests. Air is scanned in the absorbance mode for 10 min. peak-to-peak noise is recorded at 500 nm. The root mean square (RMS) noise is then calculated. The RMS noise measurement is a measure of the standard deviation of the background signals. Modern spectrophotometers are usually equipped with the noise estimation function. For older spectrophotometers, the RMS noise can be estimated by multiplying the highest peak-to-peak noise level by a factor of 0.7 (Figure 10.8).

Acceptance. The RMS noise should typically be less than 0.001 AU.

10.2.5 Baseline Flatness

The intensity of radiation coming from the light source varies over the entire UV–Vis range. Most UV–Vis spectrophotometers have dual light sources. A

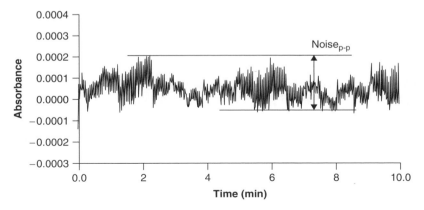

Figure 10.8. Noise measurement.

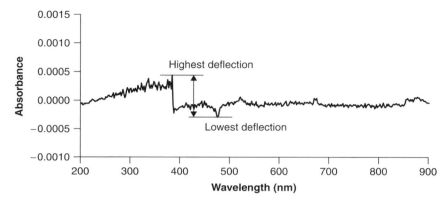

Figure 10.9. Baseline flatness.

deuterium lamp is used for the UV range and a tungsten lamp is used for the visible range. The response of the detector also varies over the spectral range. The flat baseline test demonstrates the ability of the instrument to normalize the light intensity measurement and the spectral output at different wavelengths throughout the spectral range.

Tests. Air is scanned in the absorbance mode. The highest and lowest deflections in the absorbance unit are recorded (Figure 10.9).

Acceptance. The deflection is typically less than 0.01 AU.

10.2.6 Stability

Variations in lamp intensity and electronic output between the measurements of the reference and the sample result in instrument drift. The lamp intensity is a function of the age of the lamp, temperature fluctuation, and wavelength of the measurement. These changes can lead to errors in the value of the measurements, especially over an extended period of time. The resulting error in the measurement may be positive or negative. The stability test checks the ability of the instrument to maintain a steady state over time so that the effect of the drift on the accuracy of the measurements is insignificant.

Tests. Air is scanned in the absorbance mode for 60 min at a specific wavelength (typically, 340 nm). The highest and lowest deflections in the absorbance unit are recorded (Figure 10.10).

Acceptance. The deflection is typically less than 0.002 AU/h.

10.2.7 Photometric Accuracy

The majority of quantitative applications using UV–Vis involve measurement of the standards and samples of comparable concentrations in rapid succession using

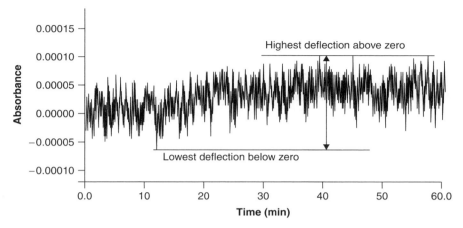

Figure 10.10. System stability.

the same instrument. As long as the photometric measurements are reproducible and the response is linear over a defined range, the absolute photometric accuracy is not critical. However, photometric accuracy is very important in the accurate measurement of the extinction coefficient that is used to characterize an analyte and to ensure that transmittance and absorbance data are comparable between spectrometers in different laboratories.

Photometric accuracy is determined by comparing the difference between the measured absorbance of the reference standard materials and the established standard value. Many solid and liquid standards are commonly used to verify the photometric accuracy of a spectrophotometer. An optically neutral material with little wavelength dependency for its transmittance/absorbance is desirable because it eliminates the spectral bandwidth dependency of measurements. The advantages and disadvantages of various commonly used photometric accuracy standards are summarized in Table 10.6. Even for a relatively stable reference standard, the intrinsic optical properties may change over time. Recertification at regular intervals is required to ensure that the certified values of the standards are meaningful and accurate for the intended use.

Tests. Either the neutral-density filters or potassium dichromate solutions are used.

Neutral-Density Filters. The empty reference filter holder (air reference) and then filters of various transmittance values at 440, 465, 546.1, 590, and 635 nm are scanned (these wavelengths are selected to minimize interaction between the absorbance and wavelength scales of the spectrophotometer being tested [12]). Compare the results with the values in the certificates. The certified values for the National Institute of Standards and Technology (NIST) SRM 930-e series glass

Table 10.6. Comparison of Some Commonly Used Standards for Photometric Accuracy Measurement

Photometric Accuracy Standard	Advantages	Disadvantages
Neutral-density glass filters NIST SRM 930 series glass filters (440–635 nm, 10%T, 20%T, 30%T) NIST SRM 1930 series glass filters (440–635 nm, 1%T, 3%T, 50%T)	Easy to handle Optically neutral; not restricted by the bandwidth Available for a range of transmittance/absorbance NIST SRM available	Limited to visible wavelengths Limited range Optical characteristics of the solid filter are different from typical samples, which are usually in solution form. Recertification required
Metal on quartz filters NIST SRM 2301 metal on quartz filter (250–635 nm, 10%T, 30%T, 90%T)	Easy to handle Optically neutral; not restricted by the bandwidth Available for a range of transmittance/absorbance Cover both UV and visible range NIST SRM available	Possible interreflections between optical surface Optical characteristics of the solid filter are different from the samples, which are usually in solution form Recertification required
Potassium dichromate in acidic solution NIST SRM 1935	Cover both UV and visible range	Relatively short stability Result may be bandwidth dependent For instrument with effective bandpass of 2 nm or less Wavelength dependent (λ_{max} and λ_{min} are measured) Careful preparation is required for precision and accuracy Corrosive and oxidative Results are sensitive to the analytical technique used

filters are listed in Table 10.7. Typical spectra of the 930-e series glass filters are shown in Figure 10.11.

Potassium Dichromate Solutions. Preparation of the potassium chromate solution differs for the NIST and BP/EP procedures. For the NIST procedure, a solution of potassium dichromate (0.06 g of dichromate per kilogram of solvent) is prepared in 0.001 N perchloric acid. A 0.001 N perchloric acid solution as the reference is scanned and then the potassium dichromate solutions at 235, 257, 313, and

Table 10.7. Standard Values for the NIST SRM 930-e Series Glass Filters (10%, 20%, and 30% Transmittance)

SRM 930-e Filter Number	Absorbance Wavelength (nm)					Transmittance Wavelength (nm)				
Set Identification	440	465	546.1	590	635	440	465	546.1	590	635
10-2106	1.082	1.006	1.034	1.007	1.028	0.0828	0.0986	0.0924	0.0838	0.0939
20-2106	0.782	0.728	0.748	0.778	0.742	0.1652	0.187	0.1786	0.1667	0.1809
30-2106	0.556	0.513	0.538	0.583	0.568	0.2778	0.3069	0.2898	0.2611	0.2704

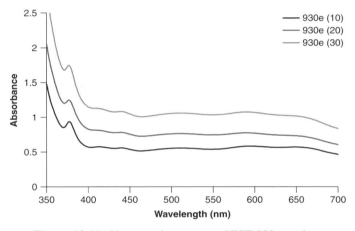

Figure 10.11. Photometric accuracy: NIST 930-e series.

Table 10.8. Standard Values for the NIST SRM 935a (Potassium Dichromate)

Wavelength (nm)	Absorbance (AU)	A (kg/g · cm)
235	0.741	12.347
257	0.862	14.374
313	0.289	4.816
350	0.642	10.692

350 nm. The standard values for the NIST potassium dichromate solutions are give in Table 10.8.

For the BP/EP procedure, a 0.006% w/v solution of potassium dichromate (60.06 mg/L) is prepared in 0.005 M sulfuric acid. A 0.005 M sulfuric acid solution as the reference is scanned and then the potassium dichromate solutions

Table 10.9. Standard Values for the BP/EP Potassium Dichromate Solution

Wavelength (nm)	Absorbance (AU)	A (1%, 1 cm)	Tolerance (A)
235	0.748	124.5	122.9–126.2
257	0.865	144	142.4–145.7
313	0.292	48.6	47.0–50.3
350	0.640	106.6	104.9–108.2

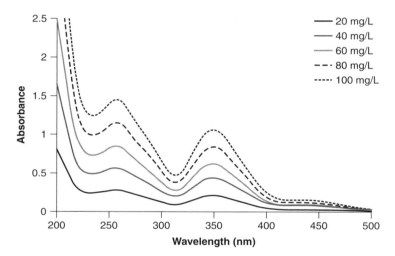

Figure 10.12. Photometric accuracy: potassium dichromate solutions.

at 235, 257, 313, and 350 nm. The standard values for the BP/EP potassium dichromate solutions are give in Table 10.9. Typical spectra of the potassium dichromate solutions of various concentrations are shown in Figure 10.12. For photometric reproducibility evaluation, the measurement at 235, 257, 313, and 350 nm is repeated six times and the % RSD of the absorbance values are calculated at each wavelength.

To alleviate the burden of trying to prepare a potassium dichromate solution of exact concentration, the specific absorbance value $A_{1\%, 1 \text{ cm}}$ can be used to compare the measured and standard values. The specific absorbance value normalizes the absorbance value obtained from using a solution of approximate 0.006% w/v to the equivalence of 1% concentration and a 1-cm path length measurement:

$$\text{specific absorbance } A_{1\%, 1 \text{ cm}} = \frac{\text{absorbance value of the standard solution prepared}}{cd}$$

where c is the % (w/v) and d is the path length (cm). For example, the measured absorbance value of 0.751 AU for a 0.0061% w/v solution will have a specific absorbance $A_{1\%,1\ cm}$ of 123.11.

Acceptance. Typical accuracy ±0.01 AU, or the specific absorbance value falls with the limit outlined in Table 10.9. Six replicate measurements of the 0.006% w/v potassium dichromate solution at 235, 257, 313, and 350 nm should be less than 0.5% RSD.

10.2.8   Linearity

The linear dynamic range of the measurement is limited by stray light at high absorbance and by noise at low absorbance. For routine measurements involving samples and the related reference chemical standards, the accuracy of the quantification of the sample depends on the precision and linearity of the measurements.

Tests. A series of potassium dichromate solutions of concentration 20, 40, 60, 80, and 100 mg/L in 0.005 *M* sulfuric acid can be used to test the linearity of the system. First, 0.005 *M* sulfuric acid as reference is scanned and then the potassium dichromate solutions at 235, 257, 313, and 350 nm. The absorbance values at various wavelength are plotted against the concentration of the solutions and the correlation coefficients are calculated.

Acceptance. Correlation coefficient $r \geq 0.999$.

10.3   PRACTICAL TIPS IN UV–VIS PERFORMANCE VERIFICATION

1. Check the intensity and the age of the lamps before carrying out the performance tests. Lamps that have been in use for a long time usually have a high noise level and unstable energy output.

2. Since photometric accuracy of the instrument depends on almost all of the performance attributes discussed above, it is better to perform the photometric accuracy test at the final stage of the regular performance verification. Once the factors that may affect the photometric accuracy have been verified, there is a better chance of passing the photometric accuracy test. There is no point in doing the photometric accuracy test if there are known problems in the other performance attributes.

3. Check the cell for the performance measurement: The quality of the measurement depends strongly on the quality of the cell used. Always use the highest-quality cell available. Make sure that the cell is clean and free of contamination. Contamination can change the absorbance of standard solutions, resulting in failure of the photometric accuracy test. The optical quality of the cell is affected by factors such as closeness to the specified path length, parallelism of the inner and outer faces of the windows, and flatness of the window. Upon

identifying a good cell that can provide accurate measurements, reserve it for calibration tests only. Use the same cell for reference and standard measurements. The same face should be facing the incident beam for the measurements. If it is not possible to use the same cell for the measurements, the optical characteristics of the cells to be used should match closely.

4. Standard reference material (SRM) for wavelength accuracy, stray light, resolution check, and photometric accuracy can be purchased from NIST. Certified reference materials (CRMs) which are traceable to NIST and recertification services can be purchased from instrument manufacturers and commercial vendors [12]. The cost of neutral-density filters and prefabricated standard solutions in sealed cuvettes can be substantial. When purchasing performance verification standards from a secondary supplier other than a national standard organizations such as NIST in the United States and National Physical Laboratory (NPL) in the United Kingdom, make sure that the traceability of the standards are available in the certificates. The traceability establishes the relationship of individual results to the national standard through an unbroken chain of comparisons.

5. The measurement of stray light at 200 nm using a potassium chloride solution can be very challenging, especially for spectrophotometers with a fixed spectral bandwidth greater than 2 nm. The measurement is taken at a wavelength very close to the rising slope of the absorption profile of the potassium chloride solution. A typical KCl solution spectrum is shown in Figure 10.5. The measurement is very sensitive to the wavelength accuracy, purity of the KCl, spectral bandwidth, and temperature [7,8]. Contrary to the general expectation that a purer standard would give better results, KCl of very high purity may fail the test for diode array spectrophotometers. A trace amount of bromide in the KCl shifts the cutoff to a longer wavelength and reduces the transmittance at 200 nm. The *European Pharmacopoeia* allows up to 0.1% bromide in the KCl solution. Band broadening due to insufficient resolution may also affect the measurement.

6. When using potassium dichromate in a sulfuric acid solution for photometric accuracy measurements, the ambient temperature should be between 21 and 23°C. For measurement at a temperature outside the range, the data should be corrected to the temperature of certification, using instructions provided in the certificate.

7. When using optical filters for photometric accuracy measurements, make sure that the filter is placed properly in the sample holder so that the filter is perpendicular to the incident light beam. Problems of reflection from the highly polished surface of the glass filter and interreflection between optical surfaces for metal on quartz filters may occur that affect the readings.

8. In case there is a need to perform wavelength accuracy and photometric accuracy measurements for the far-UV region below 240 nm, there are new certified reference standards available from Starna Cell [18]. The wavelength standard is a solution of rare earth oxides solvated in dilute sulfuric acid. The standard exhibits well-characterized absorption bands at 210, 211, 222, 240, and 253 nm (Figure 10.13). The photometric accuracy standard consists of a series

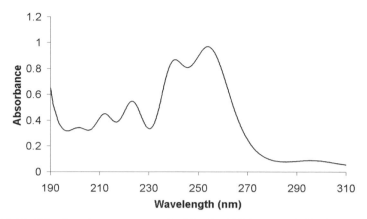

Figure 10.13. Wavelength accuracy standard for far UV. Rare earth liquid, 10-nm SBW.

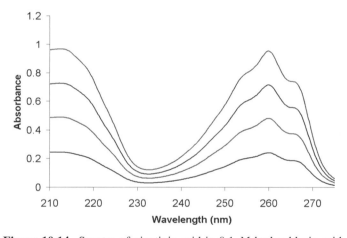

Figure 10.14. Spectra of nicotinic acid in 0.1 M hydrochloric acid.

of nicotinic acid solution in 0.1 M hydrochloric acid with concentration ranging from 5 to 25 mg/L. The absorbance of nicotine acid standards at 210 and 260 nm is shown in Figure 10.14.

10.4 CONCLUSIONS

To guarantee the reliability of results, many performance verification tests must be done on a UV–Vis spectrophotometer on a regular basis. A general understanding of the effects of each performance attribute on the outcome of the measurements will help to improve the UV–Vis experimentation.

REFERENCES

1. *United State Pharmacopoeia*, USP 24, Chapter ⟨851⟩, Spectrometry and Light-Scattering, 2000.

2. *British Pharmacopoeia*, Appendix II, B, Ultraviolet and Visible Absorption Spectrophotometry, 2001.

3. *European Pharmacopoeia*, 4th ed., 2.2.25, B, Absorption Spectrophotometry, Ultraviolet and Visible, 2002.

4. *Japanese Pharmacopoeia*, XIV, 64, Ultraviolet–Visible Spectrophotometry, 2001.

5. C. Burgess and T. Frost, eds., *Standards and Best Practice in Absorption Spectrometry*, Blackwell Science, Malden, MA, 1999.

6. T. Owen, Qualification of UV-Vis spectrophotometers: practical implementation, *BioPharm*, p. 62, Jan. 1999.

7. A. Shultz, D. Campell, and J. Messman, Reference materials standardization guidelines for the quality control and validation of UV–Vis spectrophotometers, *Cal Lab*, p. 27, Nov./Dec. 1998.

8. T. Owen, *Fundamentals of Modern UV–Vis Spectroscopy: A Primer*, Helwett-Packard, Palo Alto, CA, 1996.

9. *Performance the EP Stray Light Test with Potassium Chloride on UV–Vis Spectrophotometers*, Technical Note, Agilent Technologies, Publ. 5988-0945EN, 2000.

10. *Measuring the Stray Light Performance of UV–Vis Spectrophotometers*, Technical Note, Agilent Technologies Publ. 12-5965-9503E, 1997.

11. *Standard Test Method for Estimating Stray Radiant Power Ratio of Spectrophotometers by Opaque Filter Method*, ASTM E 387-84, 1998.

12. J. C. Travis and G. W. Kramer, NIST-traceable-reference-material optical filters program for chemical spectrophotometry, *Spectroscopy*, **14**(2), 1999.

13. *Certification and Use of Acidic Potassium Dichromate Solution as an Ultraviolet Absorbance Standard*, SRM 935, NIST Spec. Publ. 260-54, 1977.

14. *Holmium Oxide Solution Wavelength Standard from 240–640 nm*, SRM 2034, NIST Spec. Publ. 260-102, 1986.

15. *Glass Filters as a Standard Reference Material for Spectrophotometry: Selection, Preparation, Certification and Use of SRM 930 and 1930*, NIST Spec. Publ. 260-116, 1994.

16. *Standard Practice for Describing and Measuring Performance of Ultraviolet, Visible, and Near-Infrared Spectrophotometers*, ASTM E 257-93, 1998.

17. *Standard Practice for Periodic Calibration of Narrow Band-pass Spectrophotometers*, ASTM E 925-83, 1998.

18. Useful Web sites: NIST standard reference materials, *http://ts.nist.gov/ts/htdocs/230/232/232.htm*; reference materials suppliers: *www.starna.com, www.agilent.com, www.varianinc.com*.

11

PERFORMANCE VERIFICATION OF HPLC

HERMAN LAM, PH.D.

GlaxoSmithKline Canada, Inc.

11.1   INTRODUCTION

High-performance liquid chromatography (HPLC) is one of the premier analytical techniques widely used in analytical laboratories. Numerous analytical HPLC analyses have been developed for pharmaceutical, chemical, food, cosmetic, and environmental applications. The popularity of HPLC analysis can be attributed to its powerful combination of separation and quantitation capabilities. HPLC instrumentation has reached a state of maturity. The majority of vendors can provide very sophisticated and highly automated systems to meet users' needs. To provide a high level of assurance that the data generated from the HPLC analysis are reliable, the performance of the HPLC system should be monitored at regular intervals. In this chapter some of the key performance attributes for a typical HPLC system (consisting of a quaternary pump, an autoinjector, a UV–Vis detector, and a temperature-controlled column compartment) are discussed [1–8].

The performance of an HPLC system can be evaluated by examining the key functions of the various modules that comprise the system, followed by holistic testing that challenges the performance of the HPLC components as an integrated system. The holistic testing is commonly referred to as the *performance qualification* (PQ). The holistic test can be as simple as running a frequently used HPLC method in the laboratory. Modular testing of various components, which is

Analytical Method Validation and Instrument Performance Verification, Edited by Chung Chow Chan, Herman Lam, Y. C. Lee, and Xue-Ming Zhang
ISBN 0-471-25953-5 Copyright © 2004 John Wiley & Sons, Inc.

Table 11.1. Performance Attributes for HPLC Modules and Test Frequency

Module	Performance Attributes	General Expectations	Frequency
Pump	Flow rate accuracy	±2% of the set flow rate	6 months
	Gradient accuracy	±1% of the step gradient composition	6 months
	Pressure test	Proper functioning check valve Pressure decay: <75 psi/min No leak from the pump	6 months
Injector	Precision	1% RSD	6 months
	Linearity	$r \geq 0.999$	12 months
	Carryover	<1%	6 months
Detector	Wavelength accuracy	±2 nm	6 months
	Linearity of response	$r \geq 0.999$	12 months
	Noise and drift	Noise: 10^{-5} AU Drift: 10^{-4} AU/h	12 months
Column compartment	Temperature accuracy	±2°C of the set temperature	6 months

often performed in the *operation qualification* (OQ), can provide detailed information about the operation of the individual components of the HPLC. It is important to do both the individual component tests and the holistic test for the system performance verification. The common performance attributes for each HPLC module, and the general expectations for each attribute, are listed in Table 11.1.

11.2   PERFORMANCE VERIFICATION PRACTICES

11.2.1   Pump Module

Flow Rate Accuracy. One of the key performance requirements for the pump module is the ability to maintain accurate and consistent flow of the mobile phase. This is necessary to provide stable and repeatable interactions between the analytes and the stationary phase [8,9]. Poor flow rate accuracy will affect the retention time and resolution of the separation. The flow-rate accuracy of the pump can be evaluated simply by calculating the time required to collect a predetermined volume of mobile phase at different flow rate settings. For example, the flow-rate accuracy at 2 mL/min can be verified by using a calibrated stopwatch to measure the time it takes to collect 25 mL of effluent from the pump into a 25-mL volumetric flask. A calibrated flow meter can be used to determine the flow rate as well. The typical acceptance of the flow rate accuracy is listed in Table 11.1. A steady backpressure may be required, depending on the requirement of the system.

Gradient Accuracy and Precision. For gradient analysis, the ability of the pump to deliver the mobile phase at various solvent strengths over time by varying the composition of the mobile phase accurately is crucial to achieving the proper chromatographic separation and reproducibility. The gradient operation precision can be assessed indirectly by monitoring the relative standard deviation in retention time of peaks in the chromatographs from repeated injections.

The accuracy and linearity of the gradient solvent delivery can be verified indirectly by monitoring the absorbance change when the composition of the two solvents from two different channels changes. High-pressure gradient runs usually involve two solvent systems. Lower-pressure gradient LC pumps are usually equipped with quaternary proportioning valves, which can handle up to four solvents. The test will be performed for two channels at a time. For example, an LC gradient pump has four channels: A, B, C, and D. Channel A is filled with a pure solvent such as methanol, while channel B is filled with a solvent containing a UV-active tracer such as caffeine (ca. 15 mg per liter of solvent). A 99.5 : 0.5 (v/v) mixture of methanol and acetone is often used as a tracer as well. The gradient profile is programmed to vary the composition of the mixture from 100% A to 100% B in a short period of time, and then changed back to 100% A in a stepwise manner (Table 11.2). The absorbance change from 100% A (baseline) to 100% B is measured and expressed as height H in the plot of absorbance versus time (Figure 11.1). As the percentage of solvent B decreases in the solvent mixture, the UV absorbance of the mixture should decrease accordingly. If the composition of the 20% A and 80% B mixture is accurate, the height B_1, which corresponds to the absorbance at 80% B, should be close to 80% of the height H. Similarly, accuracy verifications can be determined at 60%, 40%, 20%, and 0% B. The linearity of the gradient delivery can be verified by plotting the absorbance at various mobile-phase compositions versus the theoretical composition, or simply inferred by the gradient accuracy over the range of various solvent compositions. The entire process can be repeated for channels C and D.

Table 11.2. Program for Gradient Accuracy Testing

Step	Time (min)[a]	Flow Rate (mL)	Solvent Channel A	Solvent Channel B
1	0–1	3	100	0
2	1–2	3	0	100
3	2–3	3	20	80
4	3–4	3	40	60
5	4–5	3	60	40
6	5–6	3	80	20
7	6–7	3	100	0
8	7–9	3	100	0

[a] Maintain the solvent composition.

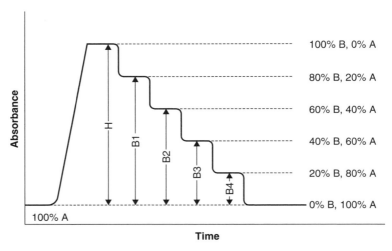

Figure 11.1. Gradient accuracy and linearity measurement.

Pressure Test. The performance of the LC pump depends on the proper functioning of the pump seal, check valves, and proper connection of the tubing. Properly functioning pump seal check valves and tubing connections are important in maintaining stable mobile-phase flow and system pressure. For pump systems that output the pressure reading in the pump head over time, a simple pressure test can be a useful qualitative test to check the performance of the seals and check valves and to determine whether or not there are any leaks in the system [9]. When performing the pump pressure test, make sure that the system is well primed so that no bubbles are present in the pump system. The pressure fluctuation caused by the presence of air bubbles in the pump will lead to misinterpretation of the test results.

The first step in the pressure test is to plug the outlet of the pump using a dead-nut. The automatic pump shutdown pressure is set to 4000 psi. The pump-head pressure signal output is connected to a recorder (this test can be performed without using a recording device by visually monitoring the pressure reading as the system is being pressurized). Pressurize the pump by pumping methanol at 1 mL/min. Methanol is less viscous than water and is more sensitive to leaks in the system. The pressure inside the pump head increases quickly as the outlet of the pump is blocked. As the pressure increases to about 3000 psi, the flow rate is reduced to 0.1 mL/min. The pressure will gradually rise to the shutdown pressure if the check valves are able to hold the mobile phase in the pump chamber as would normally be expected (Figure 11.2). If the check valve is not functioning properly, the pressure will fluctuate instead of reaching the shutdown pressure. The pressure in the pump head decreases slowly over time after the automatic shutdown. Typically, the pressure drop is less than 10% over 5 min. A steep decrease in pressure over time implies poor check-valve performance or leaks within the pumping system.

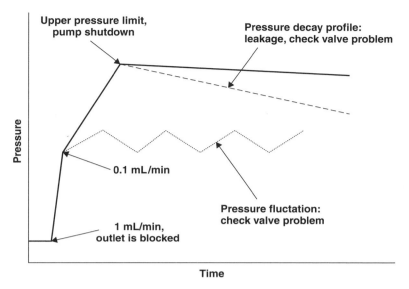

Figure 11.2. Pressure test of the pump module.

11.2.2   Injector Module

Precision. The ability of the injector to draw the same amount of sample in replicate injections is crucial to the precision and accuracy for peak-area or peak-height comparison for external standard quantitation [10,11]. If the variability of the sample and standard being injected into the column is not controlled tightly, the basic principle of external standard quantitation is seriously compromised. No meaningful comparison between the responses of the sample and the standard can be made. The absolute accuracy of the injection volume is not critical as long as the same amount of standard and sample is injected.

The volume precision of the injector can be demonstrated by making at least six replicate injections from a sample. The relative standard deviation (% RSD) of the response of the injections is then calculated to evaluate the precision. A fast and sample analysis should be considered for the test. The run condition for a very simple LC analysis of caffeine is given below as an example.

Typical HPLC Setup Parameters

- *Column:* ODS 1, 5 μm
- *Mobile phase:* water/acetonitrile (85 : 15)
- *Flow rate:* 2.0 mL/min
- *Injection volume:* 10–20 μL
- *Detection:* 272 nm
- *Temperature:* ambient

Linearity. Most automated LC injectors are capable of varying the injection volume. A variable volume of sample will be drawn into a sample injection loop by a syringe or other metering device. The uniformity of the sample loop and the ability of the metering device to draw different amounts of sample in proper proportion will affect the linearity of the injection volume. Linearity is important for methods that require the use of variable injection volumes, such as the high–low method in the quantitation of impurities (see Chapter 3 for the validation of related substances method). The linearity of the injector can be demonstrated by making injections, typically 5,10, 20, 50, and 100 μL, to cover the range 0 to 100 μL. The response of the injection is plotted against the injection volume. The correlation coefficient of the plot is used in evaluation of the injection linearity.

Carryover. Small amounts of analyte may get carried over from the previous injection and contaminate the next sample to be injected [10]. The carryover will affect the accurate quantitation of the subsequent sample. The problem is more serious when a dilute sample is injected after a concentrated sample. To avoid cross-contamination from the preceding sample injection, all the parts in the injector that come into contact with the sample (the injection loop, the injection needle, and the needle seat) have to be cleaned effectively after the injection. The carryover can be evaluated by injecting a blank after a sample that contains a high concentration of analyte. The response of the analyte found in the blank sample expressed as a percentage of the response of the concentrated sample can be used to determine the level of carryover. Caffeine can be used for the system carryover test for assessing the performance of an injector and serves as a common standard for comparing the performance of different injectors.

11.2.3 UV–Visible Detector Module

Wavelength Accuracy. *Wavelength accuracy* is defined as the deviation of the wavelength reading at an absorption or emission band from the known wavelength of the band. The detrimental effects of wavelength deviation on the qualitative and quantitative UV–Vis measurements were discussed in detail in Chapter 10. The accuracy and sensitivity of the measurement will be compromised if there is a wavelength accuracy problem. For impurities assays, the main analyte and related impurities may have a very similar UV profile. A shift in detection wavelength from the optimal setting required by the method may have a big effect on the quantitation of the impurities in relationship to the main peak [12]. The amount of impurity can be grossly over- or underestimated, depending on whether the shift is toward or away from the λ_{max} value of impurity. Inconsistency in impurity quantitation due to a problem in wavelength accuracy can complicate the method transfer process.

There are many ways to check the wavelength accuracy of a UV–Vis detector. For detectors with built-in wavelength verification, the deuterium line at 656 nm or the absorption bands at 360, 418, 453, and 536 nm in a holmium oxide filter are often used. The deuterium line and the holmium oxide bands are easy to use but are restricted to the visible range. The wavelength verification of the UV range, where most quantitative analysis is done, can be performed by filling a flow cell with a solution of a compound with a well-characterized UV absorption profile. The solution is scanned for absorption maxima and minima. The λ_{max} or λ_{min} value from the scan profile is then compared to the known λ_{max} or λ_{min} value of the compound to determine the wavelength accuracy. Solutions of potassium dichromate in perchloric acid and holmium oxide in perchloric acid can be used. However, these acidic solutions are difficult to work with, as the flow cell has to be cleaned thoroughly to remove organic residues. Residues may be oxidized by the dichromate and may alter the absorbance. The cell should be cleaned thoroughly again after the measurement to remove any traces of fluorescence from the potassium dichromate solution. Aqueous caffeine solution (12.5 mg/mL), which is easy to prepare and handle, with λ_{max} at 272 and 205 nm and λ_{min} at 244 nm, can be used. Aqueous erbium perchlorate hexahydrate (2 g in 25 mL of water, 0.14 M) solution with a λ_{max} value at 255 nm can also be used [7].

Linearity of Response. Since the analyte in the samples of interest can vary in concentration, the ability of a detector to produce a linear response to concentration variation within a reasonable range is important to the accuracy for peak area and peak height comparison between standards and samples. The linearity of the detector response can be checked by pumping or by filling the flow cell with a series of standard solutions of various concentrations. Aqueous caffeine solutions are convenient for the linearity measurement. The concentration range typically should generate responses from zero to at least 1.5 AU. Absorbencies beyond 1.5 AU are more prone to deviation due to stray light. From the plot of response versus the concentration of the solutions, the correlation coefficient between sample concentration and response can be calculated to determine the linearity (Figure 11.3a).

The linearity data can also be presented in an alternative manner by plotting the response factor (response in absorbance unit per unit concentration) versus the log of concentration (Figure 11.3b). If the responses are within the linear range, the response factors should be fairly constant. Within the linear range, the plot of response factors versus log concentration should track along a line parallel to the x-axis. A line that traces the almost constant response factors is drawn and set as 100%. Two additional lines at 95% and 105% of the respond factor values are drawn. Response factors that fall outside the region between the 95 to 105% lines are considered as nonlinear. The advantage of the response factor plot is the ability to show the upper and lower limits of the linearity range.

Noise and Drift. Electronic, pump, and photometric noise; poor lamp intensity, a dirty flow cell, and thermal instability contribute to the overall noise and drift in the detector. Excessive noise can reduce the sensitivity of the detector and hence affect the quantitation of low-level analytes [13,14]. The precision of the

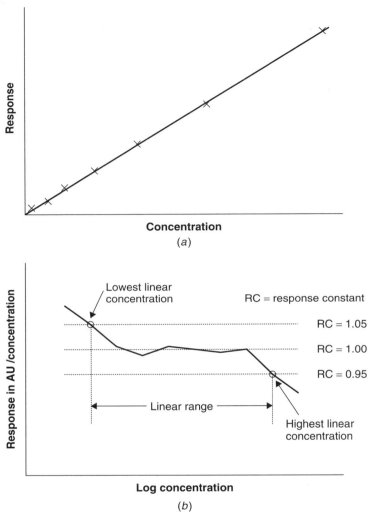

Figure 11.3. Detector linearity.

data is related to the signal-to-noise ratio. Detector drift may affect the baseline determination and peak integration. Many procedures for detector noise and drift estimation are based on ASTM (American Society for Testing and Material) Method E 685 [6,7].

Nowadays, most chromatographic software is capable of calculating the detector noise and drift. Typically, the detector should be allowed to warm up and stabilize prior to the test. Temperature fluctuations should be avoided during the test. The noise and drift tests can be performed under static and dynamic conditions. For a static testing condition, the flow cell is filled with methanol, and no flow goes through the flow cell during testing. For a dynamic testing condition, methanol is passed through the flow cell at 1 mL/min. The detector is set to 254 nm. A back pressure is required to prevent bubble formation and can be provided by attaching an appropriate size tubing to the outlet of the detector. A dummy injection with zero injection volume may be required to start the data acquisition. Noise data will be acquired for 15 min.

For short-term noise measurement, the chromatographic baseline is divided into 15 segments, each of 1-min interval (Figure 11.4). Parallel lines are drawn for each segment to enclose the peak-to-peak variation in signals. The vertical distance Y in absorbance units between the parallel lines in each segment is measured. A total of 15 Y-values will be measured. The short-term noise is estimated by summing all the Y-values and then dividing by the number of segments in the measurement. From each of the segments in the short-term noise measurement, a point in the middle of the Y-value is marked for the determination of long-term noise. There will be 15 such points in the data set.

The long-term noise can be evaluated from the same data acquisition as that for short-term noise. The time measurement is divided into two 10-min segments, from intervals of 0 to 10 min and from 5 to 15 min (Figure 11.5). Within each of the 10-min intervals, two parallel lines are drawn to enclose 10 markings in that segment. The vertical distance Z in absorbance units between the parallel lines in each segment is measured. The larger of the two Z-values provides an estimation for the long-term noise.

System drift measures the long-term shift of the baseline which is related to a systematic bias of the signal output over time [15]. Significant drift can affect proper integration of the peaks. A similar experimental setting as used for noise measurement will be used for drift measurement. The signal for the drift measurement is acquired for 60 min. The trend of the signal (which can be

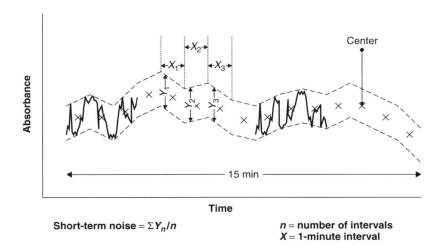

Short-term noise = $\Sigma Y_n/n$ n = number of intervals
                                        X = 1-minute interval

Figure 11.4. Short-term noise measurement.

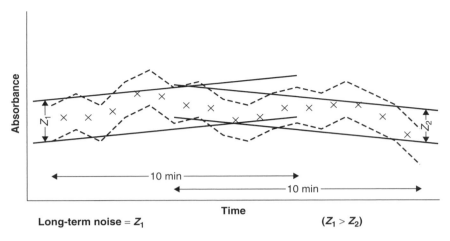

Figure 11.5. Long-term noise measurement.

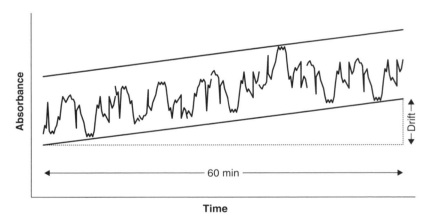

Figure 11.6. Drift measurement.

either downward or upward) is obtained by drawing a straight line to estimate the gradient of the baseline changes (Figure 11.6). The slope of the straight line expressed in absorbance units per hour can provide an estimation of detector drift.

11.2.4 Column Heating Module

The efficiency of a HPLC column varies with column temperature. In general, capacity factor k' decreases with temperature, and hence the retention of the analysis decreases with temperature [16]. Retention drops by 1 to 3% for each increase of 1°C [17]. The ability to maintain a steady and accurate column temperature is critical to achieving the desired retention time and resolution requirements in the separation consistently. Use of a column heater is a convenient way to control and maintain a steady column temperature. Even for separation performed at room temperature, the column heater can also be used to maintain a steady column temperature to prevent peak retention time shifts due to temperature variation over time due to air conditioning and heating/cooling cycles.

The temperature accuracy of the column heater is evaluated by placing a calibrated thermometer in the column compartment to measure the actual compartment temperature. The thermometer readings are compared to the preset temperature at 40 and 60°C. Many LC equipment manufacturers have the requirement set at ±2°C. The temperature stability of the column compartment can be assessed indirectly by comparing the retention time of certain peaks in the chromatogram over time. Acceptance criteria of the stability evaluation should be assessed on a case-by-case basis due to different laboratory conditions and environment.

11.3 OPERATION TIPS FOR HPC PERFORMANCE VERIFICATION

1. Routine maintenance such as pump seal replacement should be performed prior to the performance test. Check the intensity and the hours of use of the lamps before the performance tests. Lamps that have been in use for a long time usually have low intensity, a high noise level, and unstable energy output.

2. Perform the pressure test before the flow rate accuracy test to make sure that the pump seal and check valves are functioning properly and there are no leaks in the system.

3. Make sure that the flow cell is clean and free of gas bubbles when performing the detector performance tests. Dirty flow cell and gas bubbles are the main reasons for poor results for detector noise and drift.
   Typical procedure for cleaning a flow cell: The following solutions should be passed carefully through the flow cell with a syringe, taking care not to overpressurize the flow cell. Pulling the solution through the flow is usually less likely to overpressurize the flow cell.

   • 10 mL of purified lab water
   • 10 mL of methanol (HPLC grade)
   • 10 mL of purified lab water
   • 10 mL of 0.05 M NaOH, followed by 20 to 40 mL of purified lab water
   • 10 mL of 30% nitric acid followed by 20 to 40 mL of purified lab water

4. A steady pressure should be maintained within the flow cell when performing the detector linearity test. Filling the flow cell with the test solution manually with a syringe sometimes leads to pressure fluctuations and hence unstable readings. An alternative is to use the pump to deliver test solution to the flow cell. However, this option requires a large volume of test solution.

5. A stable temperature must be maintained when performing the noise and drift tests.

6. Test procedures for refractive index detectors and conductivity detectors are available from ASTM [18,19].

11.4 DISCUSSION

In reality, the performance of the LC system will deteriorate over time, especially for noise and drift. If the performance verification tests do not pass the predetermined acceptance criteria, an impact assessment should be made to evaluate the effect of the failure on the quality of the data generated by the system. The impact assessment should cover all the analyses done on the system since the last performance verification, as there is no effective way of determining when the failure occurred. The system suitability data generated together with the analyses will be very useful in demonstrating that system performance is adequate for the application at the time of analysis, so that any data generated are reliable [19].

However, running a system suitability test cannot replace the need to do instrument performance verification tests at regular intervals [20]. System suitability only demonstrates that the instrument is suitable for a particular analysis at the time of analysis. It cannot reveal marginal performance of the system. For example, the system suitability test for an HPLC assay using UV detection is unlikely to pick up any wavelength accuracy problems since both the standards and the samples are quantitated at the same wavelength. Marginal performance is an early warning of system failure. A follow-up preventive maintenance is a good way of preventing unwelcome critical system failure during an important analysis.

REFERENCES

1. D. Parroitt, Performance verification testing of high performance liquid chromatography equipment, *LC-GC*, **12**(2), 135, 1994.

2. W. Furman, T. Layoff, and R. Tetzlaff, Validation of computerized liquid chromatographic systems, *J. AOAC Int.*, **77**(5), 1314, 1994.

3. M. Dong, R. Paul, and D. Roos, Committing to calibration: a case study of one company's effort to expedite regulatory compliance, *Today's Chem.*, **10**(2), 42, 2002.

4. V.Grisanti and E. Zachowski, Operation and performance qualification, *LC-GC*, **20**(4), 355, 2002.

5. J. W. Dolan, Performance qualification of LC systems, *LC-GC*, **20**(9), 842, 2002.

6. *Standard Practice for Testing Fixed Wavelength Photometric Detectors Used in Liquid Chromatography*, Annual Book of ASTM Standards, Vol. 14.02, E 685-93, 1999.

7. *Standard Practice for Testing Variable Wavelength Photometric Detectors Used in Liquid Chromatography*, Annual Book of ASTM Standards, Vol. 14.02, E 1657-98, 1999.

8. L. R. Snyder and J. J. Kirkland, *Introduction to Modern Liquid Chromatography*, 2nd ed., Wiley, New York, 1979.

9. J. W. Dolan, Pump preventive maintenance, *LC-GC*, **15**(2), 110, 1997.

10. J. W. Dolan, Autosampler carryover, *LC-GC*, **19**(2), 164, 2001.

11. S. Kuppers, B. Renger, and V. Meyer, Autosamplers: a major uncertainty factor in HPLC analysis precision, *LC-GC Eur.*, p. 114, Feb. 2000.

12. M. Zooubair El Fallah, Performance verification: a regulatory burden or an analytical tool? *LC-GC*, **17**(5), 343, 1999.

13. D. Parriott, *A Practical Guide to HPLC Detection*, Academic Press, San Diego, CA, 1993.

14. M. D. Nelson and J. W. Dolan, UV detector noise, *LC-GC*, **17**(1), 12, 1999.

15. J. W. Dolan, Communicating with the baseline, *LC-GC*, **19**(7), 688, 2001.

16. R. G. Wolcott and J. W. Dolan, Column temperature effects in gradient elution, *LC-GC*, **16**(12), 1080, 1998.

17. J. W. Dolan, The important of temperature, *LC-GC*, **20**(6), 524, 2002.

18. *Practice for Refractive Index Detectors Used in Liquid Chromatography*, Annual Book of ASTM Standards, Vol. 14.02, E 1030-95, 1999.

19. *Practice for Testing Conductivity Detectors Used in Liquid and Ion Chromatography*, Annual Book of ASTM Standards, Vol. 14.02, E 1511-93, 1999.

20. W. Furman, J. Dorsey, and L. Snyder, System suitability tests in regulatory LC and GC methods: adjustment vs. modification, *Pharm. Technol.*, p. 58, June 1998.

12

OPERATIONAL QUALIFICATION OF A CAPILLARY ELECTROPHORESIS INSTRUMENT

NICOLE E. BARYLA, PH.D.
Eli Lilly Canada, Inc.

12.1 INTRODUCTION

It has been over a decade since the first commercial capillary electrophoresis (CE) instrument was introduced and its strengths and weaknesses identified. Its outstanding resolving ability and high efficiencies were praised. However, the instrument's robustness was less than desirable. Since then, manufacturers have addressed these concerns and have made refinements to the commercial system such that ruggedness and reproducibility have improved significantly.

Although industrial laboratories shied away from the technique at first, CE is now becoming more common in these labs for a variety of analyses, including ion analysis, chiral pharmaceutical analysis, and peptide mapping [1]. With the increased prevalence of CE in industrial analytical laboratories comes the need for instrument qualification to ensure the proper functioning and performance of the instrument in order to obtain consistent, reliable, and accurate data.

For analytical equipment, qualification is broken down into four areas: design qualification (DQ), installation qualification (IQ), operational qualification (OQ), and performance verification (PV) [2,3]. In this chapter we focus on the operational qualification of a capillary electrophoresis instrument. The tests used in the operational qualification are often used in the routine performance verification as

Analytical Method Validation and Instrument Performance Verification, Edited by Chung Chow Chan, Herman Lam, Y. C. Lee, and Xue-Ming Zhang
ISBN 0-471-25953-5 Copyright © 2004 John Wiley & Sons, Inc.

well. The recommendations for procedures and acceptance criteria outlined in this chapter are based on practical experience and are in line with the commercially available instrument OQ practices for capillary electrophoresis systems [4,5].

12.2 PARAMETERS FOR QUALIFICATION

Capillary electrophoresis is a very simple technology. A schematic diagram of a capillary electrophoresis instrument is shown in Figure 12.1. Typical capillary lengths range from 20 to 100 cm and capillary inner diameters are 10 to 100 μm. Each end of the capillary is immersed in buffer reservoirs that are connected to the high-voltage power supply (5000 to 30,000 V) via platinum electrodes. To perform a CE separation, the capillary is filled with a buffer by applying an external pressure to the inlet reservoir. Then the inlet reservoir is replaced with a sample reservoir from which a short plug of sample is pushed into the capillary by applying a small pressure. Next, the inlet reservoir is put back in place, and separation begins on the application of an electric field. Ions in the sample migrate with an electrophoretic mobility determined by their charge and mass and are detected on-capillary near the outlet reservoir using photometric detection.

Performing an operational qualification procedure ensures that the specific parts of an instrument are functioning according to defined specifications for precision, linearity, and accuracy. For operational qualification, testing individual instrument parameters and comparing them to accepted values requires isolating each parameter. Each parameter is related to a specific CE function. Typical CE functions that are subjected to qualification and their associated parameters are shown in Figure 12.2.

12.2.1 Injection Parameters

To obtain quantitative results for a particular application, it is mandatory that the instrument is capable of performing reproducible sample injections. Furthermore,

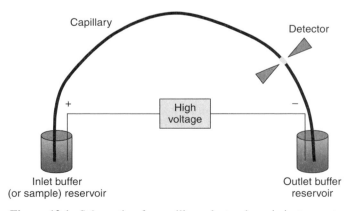

Figure 12.1. Schematic of a capillary electrophoresis instrument.

Functions   **Parameters**

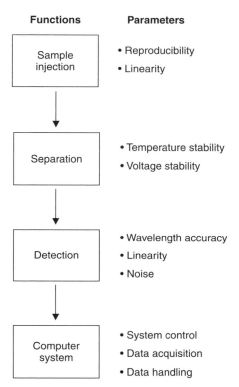

Figure 12.2. CE functions and their parameters subject to qualification testing.

evaluating the linearity of the injection mechanism is necessary to ensure that injections of varying sample volumes can be related quantitatively. Once the injection reproducibility has been demonstrated, completion of the injection linearity test infers the reproducibility at other injection volumes [4].

12.2.2 Separation Parameters

The instrumental parameters subject to qualification that are involved in the separation step are capillary thermostating and the high-voltage power supply. The heat generated by the passage of current through the capillary (Joule heat) creates radial temperature gradients, with higher temperatures at the center of the capillary than at the walls. This temperature gradient gives rise to viscosity differences in the buffer and leads to band broadening, as analytes in the centre of the capillary travel faster than those near the walls. There is also a strong viscosity dependence on sample injection and migration time; thus effective control of the capillary temperature is important for reproducible operation. During a separation, a dc power supply is used to apply up to 30,000 V (either positive or negative) across a capillary. A stable regulation of the voltage is required to obtain migration time reproducibility.

12.2.3 Detection Parameters

There are numerous methods of detection that are employed for capillary electrophoresis, such as UV absorbance, laser-induced fluorescence, electrochemical, and mass spectrometric detection. However, UV absorbance detection is the most commonly employed detection method in capillary electrophoresis since most analytes absorb light in the UV region (at least to some extent) without needing chemical derivatization. Diode array detection is an alternative to single-wavelength UV detection that provides spectral information for each analyte. The qualification parameters associated with diode array detection are wavelength accuracy, response linearity, and detector noise. These parameters are important in determining overall accuracy of results, determining accuracy of results over a range of sample concentrations, and determining the instrument's detection limits, respectively.

12.2.4 Computer System

An appropriate computer system is needed to provide system control, data acquisition, and accurate and precise integration and quantification of electrophoretic peaks. Generally, it is accepted that proper operation is confirmed by obtaining satisfactory results from operational qualification testing of the complete computer system [3]. The qualification of the computer system is beyond the scope of this chapter and is discussed in detail in Chapter 17.

12.3 QUALIFICATION PRACTICES

The frequency of operational qualification of a CE instrument depends on factors such as the manufacturer's recommended OQ interval, the required instrument performance, and the instrument's use. Note that preventive maintenance procedures such as lamp replacement may require repeating the qualification of the particular CE function [3]. The following sections describe the tests to perform and their associated limits that are intended to qualify a CE instrument for its general purpose. Troubleshooting tips for instruments that do not perform to the intended specifications are presented in Section 12.4.

12.3.1 Selection of Buffers and Test Samples

The selection of the electrophoretic buffer is critical to the success of any CE separation. The reproducibility of the migration time in a CE separation is highly dependent on maintaining a constant buffer pH during each run. Thus, the buffer selected must have good buffering capacity at the chosen pH. Generally, a buffer system has effective buffering capacity at a pH value that falls within ± 1 of a buffer's pK_a value. Furthermore, the buffer should have a low absorbance at the intended wavelength of detection. Commonly used, robust buffer systems are phosphate at pH 7 and borate at pH 9.

The test sample to use in qualification should be a small molecule that is soluble in water, has a strong UV chromophore, and possesses a charge at the intended separation pH value. A nonvolatile compound is also desirable, especially when used for assessing peak area reproducibility during the detector linearity test. Some examples of test samples that can be used are benzoic acid, 4-hydroxyphenylacetic acid, 4-hydroxybenzoic acid, and 4-hydroxyacetophenone.

12.3.2 Injection Reproducibility

To assess injection reproducibility, a series of replicate injections are performed and run using the following method. A new fused silica capillary is precon-ditioned by rinsing at high pressure (100 kPa) with 1.0 M NaOH followed by water and finally, with buffer. The temperature should be set to a constant value between 20 and 25°C. A sample (1 mM) prepared in water is injected using low pressure (5 kPa) for 3 s. A constant voltage is applied and detector set at an appropriate wavelength for the sample chosen. An example electropherogram is shown in Figure 12.3. The data rate should be set to a moderate value (5 Hz) and six replicate injections should be performed. Prior to each injection the capillary is rinsed with buffer using high pressure (100 kPa). The injection reproducibility is based on the % RSD of the corrected peak area of all six runs. The corrected peak area must be used, since in contrast to chromatographic techniques, where all analytes travel at the same rate through the column, analytes migrate with different velocities in electrophoresis. Analytes of lower mobility remain in the detection window longer than those of higher mobility, and thus show increased peak area. The corrected peak area can be calculated by dividing the peak area by the migration time. To pass the test, the % RSD should be less than 3%.

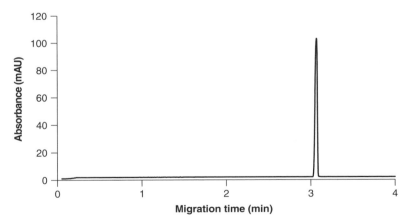

Figure 12.3. Typical electropherogram from an injection reproducibility test. CE condi-tions: capillary, 50 μm ID × 50 cm (40 cm to detector); temperature, 20°C; detection, 325 nm with 10-nm bandwidth; injection, 3 s (5 kPa) 1 mM 4-hydroxyacetophenone; applied voltage, +30 kV; separation buffer, 20 mM borate at pH 9.2; conditioning, 2-min high-pressure rinse (100 kPa) with 20 mM borate buffer pH 9.2 between runs.

12.3.3   Injection Linearity

To measure injection linearity, increasing volumes of sample are injected from a single sample vial and run using a preconditioned capillary (preconditioned as in Section 12.3.2). Prior to each injection, the capillary is rinsed with buffer at high pressure (100 kPa). Sample injections in the range 10 to 50 kPa·s are made to ascertain the linearity. Either injections using a constant pressure and varying injection times, varying pressures and constant injection times, or varying pressures and varying injection times can be used. Depending on the instrument used, the applied injection pressure may be fixed such that a constant pressure and varying injection times must be used. Examples of four injections that can be used are 5 kPa for 2 s, 5 kPa for 3 s, 5 kPa for 5 s, and 5 kPa for 10 s. After injection, a constant voltage is applied and the detector set at an appropriate wavelength for the sample chosen. The temperature is set to a constant value between 20 and 25°C, and the data rate is set to a moderate value (5 Hz). A calibration curve of corrected peak area versus injection volume is constructed and a correlation coefficient calculated. The calibration curve should be linear with an r^2 value greater than 0.99 to pass the test.

12.3.4   Temperature Stability

Most commercial CE instruments have the function to collect temperature data during an electrophoretic run such that temperature stability can be assessed. Set up a method with an initial high-pressure rinse (100 kPa) with buffer followed by application of constant voltage. Set the temperature to a constant value between 20 and 25°C and then collect temperature data for 10 min at a low data rate (1 Hz). No sample injection is necessary. The temperature should be stable within ±0.2°C.

12.3.5   Voltage Stability

As with temperature, most commercial CE instruments are able to measure fluctuations in the applied voltage over the course of a run. To assess the stability of the high-voltage source, runs should be set up to measure voltage in both positive polarity and negative polarity modes. Set up a method with an initial high-pressure rinse (100 kPa) with buffer followed by application of constant voltage (one method with +20 kV and one method with −20 kV, for example). Set the temperature to a constant value of 20 to 25°C and collect the voltage signal for 10 min at a moderate data rate (5 Hz). No sample injection is necessary. Inspect the voltage signal versus time output and determine if the mean voltage is less than 1% of the set value. If the voltage falls within these limits, the high-voltage power supply passes the test.

12.3.6   Detector Wavelength Accuracy

The wavelength accuracy of the diode array detector is commonly determined by comparing a measured absorbance with the absorbance maxima of a reference material such as a holmium oxide filter [3]. On commercial CE instruments, this test is often performed automatically by selecting a detector calibration function in the software. Prior to running the test, the lamp should be warmed up for at least 30 min. If a holmium oxide filter is used for the test (traceability of the holmium oxide filter must be documented), the spectrum should contain at least two maxima in the wavelength region of 440 to 465 nm. Alternatively, the UV spectra of a compound obtained using the diode array detector can be compared to its spectra found in a spectral library. The spectra can be extracted from the peak obtained during one of the injection reproducibility (Section 12.3.2) test runs and compared to its well-characterized spectra found in a spectral library. For example, if 4-hydroxyacetophenone was used as the test analyte for injection reproducibility, the extracted spectra from the peak should show a local maximum at 235 nm and an absolute maximum at 325 nm.

12.3.7   Detector Response Linearity

To measure detector linearity, increasing concentrations of a sample are injected and analyzed using a preconditioned capillary prepared as described in Section 12.3.2. The linearity of the detector is measured over the concentration range of 0.1 to 5.0 mM. Prior to each injection, the preconditioned capillary is rinsed with buffer at high pressure (100 kPa). Four sample concentrations that can be used are 0.1, 0.5, 1.0, and 5.0 mM. After the 3-s 5-kPa injection, a constant voltage is applied and the detector set to collect data at an appropriate wavelength for the sample. The temperature is set to a value between 20 and 25°C, and the data rate is set to a moderate value (5 Hz). A calibration curve of corrected peak area versus sample concentration is constructed and a correlation coefficient calculated. A linear response should be observed and the value of r^2 should be greater than 0.99 to pass the test.

12.3.8   Detector Noise

The detector noise can be determined by measuring the random fluctuations of the signal's amplitude over time. Set up a method with an initial high-pressure rinse (100 kPa) with buffer followed by application of constant voltage. Set the detector to collect data at a specific wavelength such as 254 nm. The test should be performed at a constant temperature between 20 and 25°C. Collect the absorbance signal for 10 min at a moderate data rate. No sample injection is necessary. The standard deviation of the signal noise is generally less than 5×10^{-2} mAU if the detector is functioning properly.

12.4   COMMON PROBLEMS AND SOLUTIONS

When performing the qualification tests, there are a few items to keep in mind when assessing the results. The tests have been presented independently for the different CE functions and their respective parameters. However, some parameters are dependent on the success of another parameter associated with a different CE function. The instrument must pass the temperature stability test prior to performing the injection reproducibility and linearity test. Since the injection depends on the viscosity of the sample solution, which in turn depends on the temperature, the temperature must remain constant in order to have reproducible and linear injections. Further, if the injection reproducibility or linearity fails, the detector linearity test may also fail, which may not necessarily be due to a detector malfunction. Rather, there could be a problem with the injection mechanism. If a parameter fails its qualification test, there are a few easy checks to perform before a service call is made.

1. *Injection parameters.* If the injection reproducibility or linearity results are problematic, ensure that the sample vial cap is put on correctly. Sometimes, if the cap is put on incorrectly, the vial cannot be pressurized and injection either fails or is irreproducible. Also, check to make sure that no air bubbles are present in the sample vial. If air is injected into the capillary, poor results will be obtained.

2. *Separation parameters.* For all electrophoretic runs, the voltage applied during the separation must be chosen such that Joule heating is prevented. One method to monitor Joule heating is to prepare a plot of the current generated with increasing voltage applied (Ohm's law plot). The linearity of the plot indicates adequate heat dissipation. The point where the plot deviates from linearity indicates that there is excessive Joule heating and that the heat dissipation of the system has been exceeded (Figure 12.4).

To keep a stable temperature and prevent Joule heating, commercial CE instruments are either liquid- or air-cooled. If the instrument is liquid-cooled, it is important to make sure that the coolant is flowing through the system. This can be checked by removing the capillary cartridge and inspecting the area of the

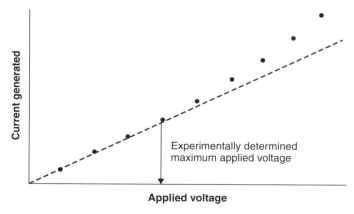

Figure 12.4. Example of an Ohm's law plot where the dashed line shows the predicted behavior if there is effective heat dissipation.

cartridge where the coolant enters and exits. If this area is damp, the coolant is flowing; if not, a service call should be made and the problem reported.

3. *Detection parameters.* If the detector is giving a noisy signal that falls outside the limits of the detector noise qualification test, determine the total lamp hours. Deuterium lamps usually have a lifetime of about 1000 h, and as a lamp ages, its output stability will deteriorate. If the lamp is getting close to its rated number of hours, or seems to take awhile to ignite (>1 min), the lamp should be changed. Furthermore, prior to performing any qualification test involving the detector signal (injection parameters and detection parameters), the lamp should be warmed up for at least 30 min.

REFERENCES

1. J. Chapman and J. Hobbs, *LC-GC*, **17**, 86–99, 1999.
2. L. Huber, *LC-GC*, **16**, 148–156, 1998.
3. V. Grisanti and E. J. Zachowski, *LC-GC*, **20**, 356–362, 2002.
4. *Agilent Capillary Electrophoresis System Operational Qualification/Performance Validation*, Agilent Technologies, Waldbronn, Germany, 2000.
5. *P/ACE™ System MDQ Operational Qualification 1*, Beckman Coulter, Inc., Fullerton, CA, Sept. 1998.

13

LC-MS INSTRUMENT CALIBRATION

Fabio Garofolo, Ph.D.
Vicuron Pharmaceuticals, Inc.

13.1 INTRODUCTION

As analytical and bioanalytical methods must be validated before using them for routine sample analysis and after changing method parameters (see Chapter 8), instruments such as liquid chromatography coupled with mass spectrometry (LC-MS) or tandem mass spectrometry (LC-MS/MS), which are utilized to perform the analysis, should be calibrated and qualified. In addition, an instrument's performance should be tested for suitability prior to use on practically a day-to-day basis.

For Good Laboratory Practice (GLP) studies, any equipment (instrument and any computer system used with it) must be tested according to the documented specifications. GLP regulations, which were issued by the U.S. Food and Drug Administration (FDA) in 1976, apply specifically to nonclinical studies used for Investigational New Drug (IND) registration. Shortly after the FDA introduced GLP regulations, the Organization for Economic Cooperation and Development (OECD) published a compilation of Good Laboratory Practices. OECD member countries have since incorporated GLP into their own legislation. In Europe, the Commission of the European Economic Community (EEC) has made efforts to harmonize the European laws. In general, instrument calibration is part of the regulatory compliance for worldwide drug submission. FDA GLP Chapter 21 of the Code of Federal Regulations Part 58 (CFR 21 Part 58) [1] deals with the

Analytical Method Validation and Instrument Performance Verification, Edited by Chung Chow Chan, Herman Lam, Y. C. Lee, and Xue-Ming Zhang
ISBN 0-471-25953-5 Copyright © 2004 John Wiley & Sons, Inc.

maintenance and calibration of equipment, including LC-MS or LC-MS/MS. The following are excerpts:

> Equipment shall be adequately inspected, cleaned, and maintained. Equipment used for the generation, measurement, or assessment of data shall be adequately *tested, calibrated* and/or *standardized*.

> The written *Standard Operating Procedures* required under §58.81(b)(11) shall set forth in sufficient detail the methods, materials, and schedules to be used in the routine inspection, cleaning, maintenance, testing, calibration, and/or standardization of equipment, and shall specify, when appropriate, remedial action to be taken in the event of failure or malfunction of equipment. The written standard operating procedures shall designate the person responsible for the performance of each operation.

> *Written records* shall be maintained of all inspection, maintenance, testing, calibrating and/or standardizing operations. These records, containing the date of the operation, shall describe whether the maintenance operations were routine and followed the written standard operating procedure.

According to the GLP, SOPs are defined as procedures that contain the details of how specified tasks are to be conducted. The GLP definition of SOP merges the International Organization for Standardization (ISO) definitions of *procedure* and *work instructions* where *procedure* is a general statement of policy that describes how, when, and by whom a task must be performed; and *work instructions* contain the specific details of how the laboratory or other operation must be conducted in particular cases [2].

In addition to GLP, any instrument used to perform analysis under Good Manufacturing Practices (GMPs), must also be covered by instrument SOPs. If a LC-MS laboratory never performs any nonclinical GLP work for FDA submission, GMP, or bioavailability/bioequivalence testing, that laboratory in theory is not obligated to have LC-MS SOPs [3]. However SOPs are very useful because they provide a measure of consistency in how data are generated, processed, and archived. This consistency has many benefits: for example, facilitating discovery of the cause of any anomalous data that may be produced. For these reasons, many LC-MS laboratories, even though not required to be in GLP compliance, have decided to operate under these regulations and use SOPs at all times. It is important to clarify that the main purpose of the LC-MS SOPs is not to substitute the operator's manuals but to ensure that a particular instrument is properly maintained and calibrated such that any data generated from it, when operating, can be considered reliable [3].

In this chapter we focus primarily on calibration of LC-MS where the mass spectrometer is operating at *unit resolution*, resolution that is sufficient to separate two peaks one mass unit apart. This kind of low-resolution mass filter covers almost 90 percent of the instruments commonly used for qualitative and/or quantitative analysis of small molecules. Batch-to-batch qualification testing of the instrument is also described. For the calibration of high-resolution mass spectrometers such as magnetic sector, TOF, or FTICR coupled with liquid chromatography, readers are referred to specific publications.

13.2 PARAMETERS FOR QUALIFICATION

The day-to-day performance of a given LC-MS or LC-MS/MS depends on its calibration, tuning, system suitability test, and final overall validation.

13.2.1 Calibration Parameters

Calibration parameters are instrument parameters whose values do not vary with the type of experiment, such as peak widths, peak shapes, mass assignment, and resolution versus sensitivity.

Peak Width. Peak width depends on the mass resolution. A resolution of 1 mass unit is sufficient to distinguish ions in most qualitative/quantitative small molecule applications. A typical definition of unit resolution is when the peak width at half-height is about 0.6 to 0.8 mass unit. The profile scan of ions on a typical benchtop LC-MS has a bandwidth of about 1 mass unit (Figure 13.1).

Peak Shape and Profile Scan. In a typical benchtop LC-MS, abundance measurements are collected at 0.10-*m/z* increments, as shown in Figure 13.2. When these data are presented in a mass spectrum, a single line can be shown. The height and position are derived from the profile scan.

Mass Assignment. It is performed using specific MS calibrants. Calibrants should be well-characterized reference materials. Certification and handling of these

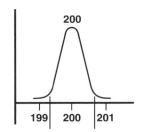

Figure 13.1. Bandwidth in a typical benchtop LC-MS.

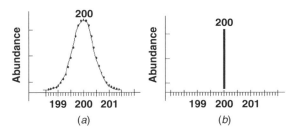

Figure 13.2. (*a*) Profile scan and (*b*) spectral representation in a typical benchtop LC-MS.

compounds should be documented. LC-MS qualitative analysis always uses reference materials, whereas quantitative analysis uses a combination of reference materials and actual analyte standards as calibrants.

Resolution versus Sensitivity. Mass resolution is a compromise between ion intensity and peak width. In general, as the resolution is increased, the ion intensity decreases (see Section 13.3.2).

Calibration. Calibration has been defined as "a comparison of a measurement standard or instrument of known accuracy with another standard or instrument to eliminate deviations by adjustment" [4]. The term *calibration* in this chapter is used to indicate a mass-axis calibration. Other calibration (e.g., voltage of power supplies) is outside the scope of this chapter. Calibration of the mass axis requires a series of ions evenly spaced throughout the mass range. When a calibration is completed, it is possible to acquire data over any mass range within the calibrated range. It is therefore sensible to calibrate over a wide mass range. Manual, semiautomatic, and automatic LC-MS calibration require introduction of the solution of the calibrant (calibration solution) into the MS at a steady rate while the procedure is running. The calibration solution is introduced directly (infused) into the MS from a syringe pump or through a loop injector connected to the LC pump. It is recommended that the MS be calibrated at least once every three months and the calibration checked about once a week.

13.2.2 Tuning Parameters

Tuning parameters are instrument parameters whose values can vary with the type of experiment. For example, if the experiment requires quantitative data on one or more particular ions, the MS should be tuned to optimize the response for the specific analyte standard. Manual, semiautomatic, and automatic tuning procedures require the introduction of a tuning solution of the analyte of interest into the MS at a steady rate. This can be done in three different ways: (1) by introducing the solution directly from the syringe pump (direct infusion); (2) by introducing the sample from the syringe pump into the effluent of the LC by using a tee union; and (3) by injecting the sample into the effluent of the LC by using a loop injection valve [flow injection analysis (FIA)].

The first method is good for tuning for experiments at a low flow rate involving the syringe pump. The second and third methods are useful for experiments at a higher flow rate involving the LC. In general, the FIA gives a better evaluation of the instrument signal-to-noise ratio (S/N) at the masses of interest. In most cases the tuning parameters obtained from the automatic or semiautomatic tuning procedures are sufficient for many analytical experiments. However, one must ensure that proper parameters are generated. For some applications it is necessary to do manual fine-tuning of several MS parameters.

The optimized parameters, which affect the signal quality, change from instrument model to instrument model and from brand to brand. Examples are: source temperature, ionization voltages, gases (nebulization, desolvation, and collision), ion path potentials (lens, multipoles, or stacked rings), collision energy, solution, or mobile-phase flow rate. The potentials, RF values, and gas pressure affect the declustering, focusing, fragmentation, and efficiency of ion transmission.

13.2.3 System Suitability Testing

System suitability allows the determination of system performance by analysis of a defined solution prior to running the analytical batch. System suitability should test the entire analytical system, chromatographic performance as well as the sensitivity of the mass spectrometer for the compounds of interest. Some LC-MS SOPs reference analytical methods as the source of operating details for a given analysis. This works particularly well for quantitative analysis, where analytical methods include critical details on instrument parameters and special calibrations that might be required for a particular analyte. Thus, system suitability testing provides the daily [3] checking of the system.

13.2.4 Validation

Validation is the final step to guarantee that a LC-MS system performs as expected. Validation includes instrument calibration, tuning, testing, and checking of the documentation for completeness, correctness, and compliance with SOPs. Validation consists of four separate steps:

1. Validation of the instrument and the computer controlling it (computer system validation or CSV)
2. Validation of the analytical method running on that equipment
3. System suitability testing, to test the equipment and the method together to confirm expected performance
4. QA/QC review of sample analysis data collected on such a system [4]

13.3 CALIBRATION PRACTICES

How often an LC-MS should be calibrated depends on the mass accuracy required. For example, instrument calibration should be verified daily when performing accurate mass measurements of peptides and proteins. However, the quantitative analysis of small molecules requires less frequent calibration.

13.3.1 General Tuning and Calibration Practice for MS

The mass analyzer should be calibrated on a regular basis by infusing a calibration solution. In general, an electrospray ionization source (ESI) is used. The solution should produce ions (with known exact masses) that cover the entire instrument mass range or at least the mass range that will be used for subsequent analyses.

For the most recent LC-MS on the market, an automatic procedure is included in the software package to tune and calibrate in the ESI mode. However, older instruments and/or very specific applications still require manual or semiautomatic procedures to optimize the parameters that affect ion detection. In an LC-MS instrument, the mass spectrometer is tuned and calibrated in three steps: (1) ion source and transmission optimization, (2) MS calibration, and (3) fine tuning (detection maximization of one or more particular ions).

Ion Source and Transmission Optimization. In this step, the MS in ESI mode is roughly tuned on one or more specific ions by infusing the calibration solution at a steady rate around 5 μL/min for several minutes. The introduction of the calibration solution compound is best achieved using a large-volume Rheodyne injector loop (50 or 100 μL) or an infusion pump (e.g., a Hamilton syringe pump): When using a large-volume injection loop, the solvent delivery system should be set up to deliver around 5 μL/min of 50 : 50 acetonitrile/water or 50 : 50 methanol/water through the injector into the source. An injection of 50 μL of calibration solution will then last for at least 10 min. When using an infusion pump, the syringe should be filled with the calibration solution and then connected to the ESI probe with fused silica or peek tubing.

The ions chosen for the optimization should be in the middle of the calibration range or in a specific region of interest (e.g., $m/z = 195$ of caffeine or $m/z = 906.7$ of PPG calibration solution). Before using the automatic tuning procedure to demonstrate that the transmission of ions into the MS is optimum, the ESI source sprayer should be manually adjusted to establish a stable spray of ions into the MS and to ensure that enough ions are detected to calibrate the MS.

The degree of adjustment changes from model to model and depends on the source geometry. In general, sources where the spray is off axis with the inlet, such as orthogonal and z-spray, require fewer adjustments. For the on-axis sources, to avoid the contamination of the instrument optics, it is important to check that the source is not spraying directly at the instrument's orifice. On-axis sources usually have a side window to provide a good view for the spray trajectory. In both cases the sprayer should always be moved in small increments until reaching the optimum position for the highest S/N value of the ions of interest. The source parameters (e.g., nebulizer gas, lens voltages) are optimized automatically, semiautomatically, or manually.

MS Calibration. In the second step, the MS in the ESI mode is calibrated using the calibration solution. All mass analyzers must be calibrated. For example, quadrapoles 1 and 3 are both calibrated for the triple-quadrupole mass spectrometer. Here is the calibration process:

1. A mass spectrum of a calibration solution is acquired (calibration file) and matched against a table of the expected masses of the peaks in the calibration solution that are stored in a reference file.
2. Each peak in the reference file is matched to a corresponding peak in the calibration file.

3. The corresponding matched peaks in the calibration file are the calibration points.
4. A calibration curve is fitted through the calibration points.
5. The vertical distance of each calibration point from the curve is calculated. This distance represents the remaining (or residual) mass difference after calibration.
6. The standard deviation of the residuals is also calculated. This number is the best single indication of the accuracy of the calibration [5].

Before starting the automated calibration, a number of the peak match parameters need to be set. These parameters determine the limits within which the acquired data must lie for the software to recognize the calibration masses and result in a successful calibration. Some of these parameters are:

1. *Peak search range.* This is the window used by the calibration software to search for the most intense peak. Increasing this window gives a greater chance of incorrect peak matching. It is important to ensure that the correct peak is located in the peak search range; otherwise, a deviation in the calibration may arise. If an incorrect peak is located, the search peak range should be adjusted to locate only the correct ion.
2. *Peak threshold.* All peaks in the acquired spectrum below the intensity threshold value (measured usually as a percentage of the most intense peak in the spectrum) will not be used in the calibration procedure.
3. *Peak maximum standard deviation or maximum difference between the predicted and actual mass.* During calibration the difference between the measured mass in the acquired calibration file and the true mass in the reference file is taken for each pair of matched peaks. If this value exceeds the set value, the calibration will fail. Reducing the value of the standard deviation gives a more stringent limit, while increasing the standard deviation means that the requirement is easier to meet, but this may allow incorrect peak matching.
4. *Peak width.* This is a measure used to set the resolution, usually specified at 50% of maximum intensity.

If the acquired spectrum looks like the reference spectrum and all of the peaks expected pass the criteria above, the calibration is acceptable. If the instrument has never been calibrated before or the previous calibration file has been misplaced, a more detailed calibration routine must be used. This calibration procedure requires the location of known peaks, starting with the lowest mass ions. When a peak has been located, a digital-to-analog conversion value is assigned to the exact mass of the ion.

Once a full instrument calibration is in place, it is not always necessary to repeat the full calibration procedure when the instrument is next used. Instead of a full calibration, a calibration verification can be performed by infusing the calibration solution and setting all peak matching parameters to the values that were used for the full calibration. If the difference between the predicted and actual mass is not significant, the original mass calibration curve is still valid. Independent of the instrument model/brand, it is always possible to obtain, at the end of the auto-mated calibration, a final report that displays the predicted exact calibration solu-tion masses, the actual mass of each peak as seen in the mass spectrum, the differ-ence between these two masses (accuracy of the calibration), and in many cases also the intensities and peak widths of all the ions specified in the experiment.

Fine Tuning. In the third step, the detection of one or more particular ions is optimized to tune the MS with the standard of the analyte of interest, if available, in either the ESI or APCI mode. The mass-to-charge ratio of the analyte of interest or, alternatively, when the standard of the analyte is not available, an ion in the calibration solution that is the closest to the mass-to-charge ratio for the ion of interest, can be chosen (see Section 13.2.2).

13.3.2 Quadrupole Mass Filter Calibration

Resolution versus Sensitivity. A quadrupole mass filter can be programmed to move through a series of RF and dc combinations. The Mathieu equation, which is used in higher mathematics, can be used to predict what parameters are necessary for ions to be stable in a quadrupole field. The Mathieu equations are solved for the acceleration of the ions in the X, Y, and Z planes. A selected mass is proportional to (dc × RF × inner radius)/(RF frequency). For a given internal quadrupole radius and radio frequency, a plot can be made of RF and dc values that predict when a given mass will be stable in a quadrupole field. This is called a *stability diagram* (Figure 13.3). RF and dc combinations follow the value shown

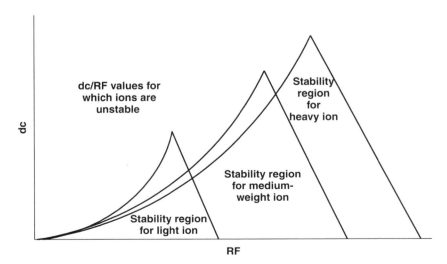

Figure 13.3. Stability diagram of a quadrupole mass filter.

by the line intersecting the stability diagram. The line that reflects the RF/dc is called a *scan line* (Figure 13.4). Modifying the scan line changes the ratio of RF to dc. The gain changes the ratio of RF to dc more rapidly at high values of RF than at low values of RF.

Constant-peak-width profile scans can be achieved by adjusting the RF-to-dc ratio. In this case the points of intersection on the scan line will be at the same distance below the apices (Figure 13.5). To adjust the scan line, the amount of

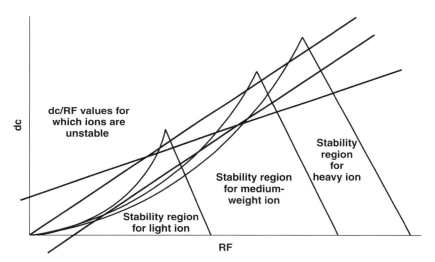

Figure 13.4. Different scan lines in a quadrupole mass filter.

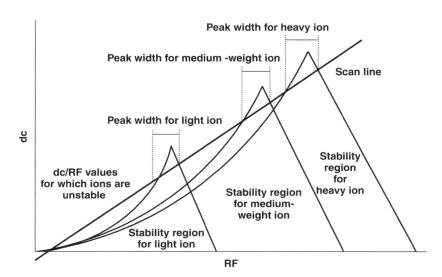

Figure 13.5. Constant peak profile scan in a quadrupole mass filter.

It is usually desirable to have a consistent peak width over the entire mass range. If the dc is held constant, the peak width varies over the mass range and increases as the mass increases (Figure 13.7). Adjusting the slope of the operating line increase the resolution. The resolution normally obtained is not sufficient to deduce the elemental analysis. Usually, quadrupole mass spectrometers are low-resolution instruments and operate at unit resolution.

The wide peak width at higher masses results in loss of resolution. In Figure 13.8 the heavy ion and its ^{13}C isotope cannot be distinguished, while

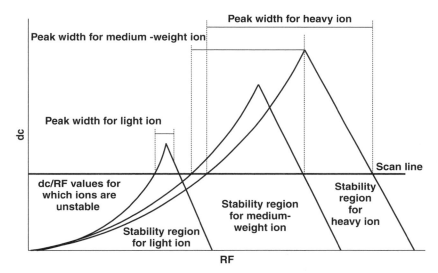

Figure 13.7. Increase of peak width when dc is held constant in a quadrupole mass filter.

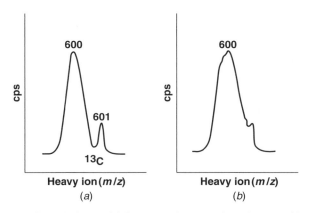

Figure 13.8. Loss of resolution at high masses in a quadrupole mass filter: (*a*) unit resolution; (*b*) low resolution.

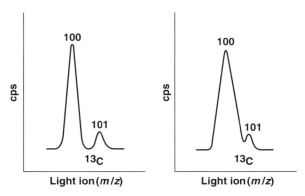

Figure 13.9. Isotope peak resolved at low masses with poor resolution at high masses in a quadrupole mass filter: (*a*) unit resolution; (*b*) low resolution at high masses.

at low masses (light ion) it is still possible to resolve the isotope peak (Figure 13.9). This is a classic case of poor resolution at high masses. The normal trade-off between high ion transmission and narrow peak widths should be optimized for each application. Most applications require unit mass resolution (i.e., isotope peaks 1 amu apart are clearly defined) which corresponds to peak widths of approximately 0.7 ± 0.1 amu at 50% intensity [full width at half maximum (FWHM)].

The entire quadrupole mass range can be divided into several mass regions such that the resolution can be adjusted in each region without affecting the others. This feature allows fine tuning of the instrument resolution for specific applications. The resolution within the individual mass regions is adjusted by increasing or decreasing the offset parameter at the masses that bracket the region. Increasing the offset increases the resolution and decreasing the offset decreases the resolution. Changing offset changes the slope of the scan line. In other words, to obtain unit mass resolution throughout the operating mass range, different scan lines are employed for different m/z ranges. For example, if the resolution of an ion at m/z 500 needs to be increased, the offset at the mass that bracket the region, for example, mass 100 and/or mass 1000, should be increased. Both the mass settings at 100 and 1000 will change the resolution, so either setting may be adjusted. However, if the resolution of an ion at m/z 999 needed to be adjusted, the offset at mass setting 1000 would have a large effect, whereas the mass setting at 100 would have a very small effect on its resolution [6,7].

Resolution is also affected by the actual time that ions spend in the quadrupole. Ions that have higher kinetic energy have shorter residence time and lower resolution. Reducing the kinetic energy typically leads to an improvement in resolution. A *dramatic change in resolution* may cause a shift in the instrument calibration. After adjusting the resolution, it is necessary to check the calibration (calibration verification), and if it is noted that the difference between the predicted and actual mass values have shifted, the instrument should be recalibrated.

Calibration Curves. A quadrupole mass spectrometer can require up to three calibration curves:

1. A *static calibration* is used to accurately "park" the quadrupole mass analyzer on a specific mass of interest. If only a static calibration is performed, the instrument is calibrated for acquisitions where the quadrupoles are held at a single mass as in SIM or SRM.
2. A *scanning calibration* enables peaks acquired in a scanning acquisition to be mass measured accurately (scan mode). If only a scanning calibration is performed, the instrument is calibrated correctly only for scanning acquisitions over the same mass range and at the same scan speed as those used for the calibration. The scan speed recommended for the scanning calibration is 100 amu/s.
3. A *scan speed compensation calibration* compensates for lag time in the system when the instrument is scanned rapidly. If only a scan speed compensation is performed (without a scanning calibration having been performed), the scan speed compensation is treated as a scanning calibration and the instrument is calibrated correctly only for scanning acquisitions over the same mass range and at the same scan speed as those used for the calibration. The scan speed recommended for the scan speed compensation is 1000 amu/s.

For some MS models and brands it is recommended that all three types of calibration are performed so that any mode of data acquisition can be used and mass ranges and scan speeds can be changed while maintaining correct mass assignment [5]. Some other instruments calibrate all together without distinctions [6–8]. When the instrument is fully calibrated, any mass range or scan speed is allowed within the upper and lower limits dictated by the calibrations.

When each peak in the reference spectrum has been matched with a corresponding peak in the spectrum acquired, the mass difference is calculated for each pair of peaks (see Section 3.1.2). These mass differences are plotted as points on a graph; each data point has the mass of the acquired peak as its x coordinate, and the mass difference above as its y coordinate, and a smooth curve is drawn through the points (Figure 13.10) [5]. The polynomial order parameter controls the type of curve that is drawn and can be set to any value between 1 and 5:

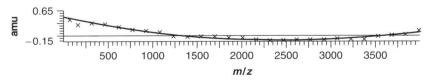

Figure 13.10. Plot of the mass difference between reference spectrum and acquired spectrum. MS1 static calibration, 28 matches of 28 tested references, SD = 0.0465. (Courtesy of Micromass Quattro LC.)

- *Polynomial order = 1*. A straight line is drawn through the points. This polynomial of order 1 is not used to calibrate quadrupole MS.
- *Polynomial order = 2*. A quadratic curve is drawn through the points. An order of 2 is suitable for wide mass ranges at the high end of the mass scale and for calibrating with widely spaced reference peaks. This is the recommended polynomial order for calibration with sodium iodide (see Section 13.3.3), which has widely spaced peaks (150 amu apart) and horse heart myoglobin (see Section 13.3.3), which is used to calibrate higher up the mass scale. In general, for a typical ESI calibration where the mass range calibrated is from 600 amu to greater than 1000 amu, the recommended setting for the polynomial order parameter is 2.
- *Polynomial order = 3*. A cubic curve is used.
- *Polynomial order = 4*. Used for calibrations that include the lower end of the mass scale, with closely spaced reference peaks. This is suitable for calibrations with polyethylene and poly propylene glycols (see Section 13.3.3) that extend below 300 amu.
- *Polynomial order = 5*. Rarely has any benefit over a fourth order fit.

For more information about the calibration procedure for a specific model or brand quadrupole MS, it is suggested that the reader consult the operators manual provided with the instrument.

13.3.3 Calibration Solutions

Calibrants are required to calibrate the mass scale of any mass spectrometer, and it is important to find reference compounds that are compatible with a particular ion source. Calibrants commonly used in electron ionization (EI) and chemical ionization (CI), such as perfluorocarbons, are not applicable in the ESI mode. The right calibrants for LC-ESI-MS should (1) not give memory effects; (2) not cause source contamination through the introduction of nonvolatile material; (3) be applicable in both positive- and negative-ion mode. The main calibrants used or still in use to calibrate ESI-MS can be divided into the following categories: polymers, perfluoroalkyl triazines, proteins, alkali metal salt clusters, polyethers, water clusters, and acetate salts.

Polymers. Polymers, such as polypropylene glycols (PPGs) and polyethylene glycols (PEGs) are the preferred calibrant for many small molecule applications. PPG calibration solutions produce mostly singly charged ions over the entire instrument mass range in both positive- and negative-ion mode. The PPG ions in general used for calibration in positive mode are: 59.0 (solvent and fragment ion), 175.1 (fragment ion), 616.5, 906.7, 1254.9, 1545.1, 2010.5, and 2242.6 (Figure 13.11) [6,7,10,11,27]. During the calibration procedure, these ions are listed in the reference file, and they should be the most intense peaks within the search range in the calibration file around the predicted mass. It is important to ensure that the correct peaks are located in the search range; otherwise, a

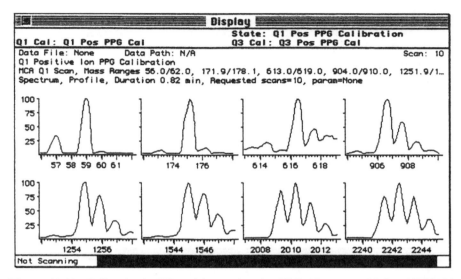

Figure 13.11. Calibration in ESI positive mode using PPG. (Courtesy of Sciex API3000.)

deviation in the calibration may arise. If an incorrect peak is located (typically, a peak with a mass difference greater than 1 amu from the predicted mass), the search range or the experiment scan width should be adjusted to locate only the correct ion. When the calibration is done using PPG, for most applications a difference between the predicted and actual mass of 0.05 amu or less can be considered as not significant. In general, a positive-ion calibration will be sufficient for the negative-ion mode, but occasionally, there might be very small calibration shifts in part of the mass range (e.g., 0.1 amu offset at low mass). The PPG ions used for calibration in negative-ion mode are 45.0 (solvent ion), 585.4, 933.6, 1223.8, 1572.1, 1863.3, 2037.4, and 2211.6 (Figure 13.12) [6,7].

PPG and PEG calibration solutions are used most widely for ESI-MS calibration, although significant source contamination and memory effects may occur [5,9,10,11,27]. Some precautions should be taken when using PEG or PPG as calibrant. Indeed, if a very low threshold and wide peak search range is used, it may be possible to select the wrong peaks and get a "successful wrong" calibration. Therefore, caution should be used when calibrating with PEG or PPG in ESI mode due to the number of peaks that are produced. Although ammonium acetate is added to the calibration solution to produce mainly $[M + NH_4]^+$ ions, under some conditions it is quite usual to see $[M + H]^+$, $[M + Na]^+$, and doubly charged ions.

The spectrum shown in Figure 13.13 demonstrates how a PEG spectrum can be dominated by doubly charged ions ($[M + 2\ NH_4]^{2+}$) if the wrong conditions are chosen. In this case the concentration of ammonium acetate in the reference solution was too high (5 mM ammonium acetate is the maximum that should be used) and the declustering potential is too low. Doubly charged ions can be

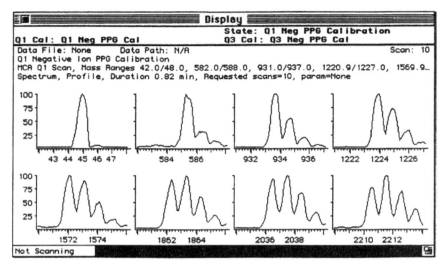

Figure 13.12. Calibration in ESI negative using PPG. (Courtesy of Sciex API3000.)

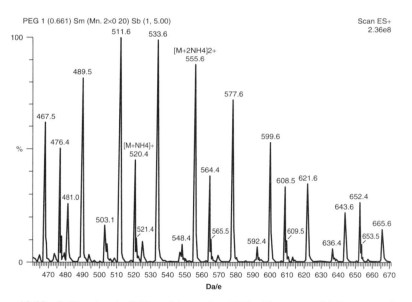

Figure 13.13. Calibration in ESI positive using PEG. (Courtesy of Micromass Quattro LC.)

identified easily because the peaks in a given cluster (e.g., one that for all $n^{12}C$ atoms and its neighbor that contains one ^{13}C and $n - 1^{12}C$) are separated by 0.5-m/z units (Thomson) instead of 1.0-m/z units. If the instrument is set to unit mass and data are acquired in continuum mode, the doubly charged peaks will appear broader, as the isotopes will not be resolved. With PEG the possible calibration range depends on the molecular weight distribution of the PEGs used in the reference solution.

Perfluoroalkyl Triazines. Perfluoroalkyl triazines such as Ultramark 1621 have mainly been used as the calibrant for FAB [10–13]. Ultramark 1621 is also used as a calibration solution for ESI-MS calibration [8,10,11,14]. Although effective, this standard is very "sticky" and is very difficult to remove from the ion source. For this reason, Ultramark 1621 calibration solution should not be used at flow rates above 10 μL/min, to avoid system contamination. Ultramark 1621 is in general used together with other calibrants to cover the entire MS range. Calibration solutions containing caffeine, L-methionyl-arginyol-phenylalanylalanine acetate H_2O (MRFA) and Ultramark 1621 are commonly employed for ESI-MS calibration [8]. It is possible to observe the following singly charged, positive ions for caffeine, MRFA, and Ultramark 1621 (Figure 13.14): caffeine: m/z 195; MRFA: m/z 524; and Ultramark 1621: m/z 1022, 1122, 1222, 1322, 1422, 1522, 1622, 1722, 1822.

Proteins. ESI-MS calibration was initially performed using solutions of gramicidin S, cytochrome c, or myoglobin as calibrant. Proteins produce multiply charged species in ESI. When proteins are used as calibrants for low-resolution LC-MS and LC-MS/MS instruments, it is not possible to resolve individual isotope peaks, and the calibration should be performed by using the average molecular mass for the unresolved isotope clusters. The danger in using proteins as calibrants is illustrated by the fact that equine myoglobin was used as a reference standard by many groups until it was discovered [10,11,15,16] that the amino acid sequence used to calculate the molecular mass of myoglobin was incorrect. The correct molecular mass of equine myoglobin was actually 1 amu higher than the value used as a reference standard.

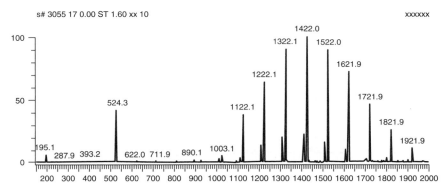

Figure 13.14. Calibration in ESI positive using caffeine, MRFA, and Ultramark. (Courtesy of ThermoFinnigan LCQ Deca.)

Alkali Metal Salt Clusters. Alkali metal salt clusters cover a wide m/z range and are used to calibrate mass spectrometers in ESI mode. Cesium iodide solutions produce singly and doubly charged species from m/z 133 up to m/z 3510 or higher [10,11,17]. Unfortunately, cesium iodide calibration solutions are not very commonly used, due to the following drawbacks: (1) sample suppression; (2) persistence in the ESI source; (3) possible cation attachment; and (4) the large spacing, 260 amu, between peaks. A mixture of sodium iodide and rubidium iodide calibration solution is able to cover the instrument's full mass range from 23 to 3920 (Figure 13.15) [5,10,11,26]. The peak at 23 is sodium, the 85 peak is rubidium, and the others are clusters. A mixture of ammonium acetate, tetrabutylammonium iodide, benzyltriphenylphosphonium chloride, and hexadecylpyridinium chloride was used successfully and described in the literature [18].

Polyethers. Polyethers such as polyethylene oxide (PEO) and polypropylene oxide (PPO) have been used for ESI-MS calibration [10,11,19]. The predominant ions for these calibrants are cation attachments, and sodium attachment is frequently observed, due to traces of sodium in solvents and glassware. The positive-ion ESI mass spectra of PEO and PPO are characterized by abundant $[M + n\mathrm{Na}]n+$ and some $[M + n\mathrm{H}]^{n+}$ species. Macrocyclic polyethers and crown ethers were also used as ESI-MS calibrants [11]. In general, nonderivatized polyethers show the following drawbacks when used as calibrations solutions: (1) they are difficult to flush out of the ion source, (2) they generate complex mass spectra resulting from the presence of several different cation sources, and (3) they have different charge states. Negative-ion ESI-MS show relatively weak $[M - H]$ peaks that can be observed only with difficulty; thus, polyethers are not useful calibration compounds for negative-ion ESI analysis.

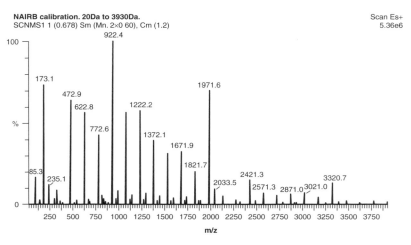

Figure 13.15. Calibration in ESI positive using sodium iodide and rubidium iodide. (Courtesy of Micromass Quattro LC.)

Derivatized polyethers such as polyether sulfate have been investigated for both positive- and negative-ion calibration [11]. Although polyether sulfates are not commercially available, they are easily synthesized. Lauryl sulfate ethoxylates were also used as calibrants for negative-ion ESI. Polyether amines and quaternary ammonium salts were used as positive-ion calibration solutions [11]. These commercially available compounds do not exhibit significant sodium or potassium adducts, and they are more easily flushed out of the mass spectrometer ion source than are nonderivatized polyethers. In addition, doubly charged polyether diamines can produce reference peaks at low m/z values.

Water Cluster. Numerous groups have used water clusters successfully as calibration solutions [10,11,20–22]. Water clusters do not produce any source contamination in ESI-MS and provide closely spaced reference peaks with a calibration range up to m/z 1000. In positive-ion mode, protonated water clusters with up to 70 water molecules are observed. In negative ESI singly deprotonated water clusters are observed [$OH^- \cdot (H_2O)_n$ with $n > 20$], as well as solvated electrons [$(H_2O)_m^-$ with $m > 11$].

Acetate Salts. Sodium acetate and sodium trifluoracetate clusters were used and produce useful reference peaks for both positive and negative ESI [10,11,23]; 0.5% acetic acid in ammonium acetate solutions can be used for calibration in ESI-MS. This calibration solution, which is volatile, produces cluster ions up to m/z 1000. Therefore, it does not produce any source contamination or memory effects. Replacing acetic acid by trifluoroacetic acid (TFA) further enlarges the mass range to m/z 4000, but TFA produces some memory effects and ion suppression, especially in negative-ion mode.

Other Calibrants. The use of cesium salts of monobutyl phthalate and several perfluorinated acids to generate cluster ions up to m/z 10,000 has been described in the literature [11,24]. Fluorinated derivatives of glyphosate and aminomethylphosphonic acid were used as reference compounds for negative ESI [11,20]. These calibrants were synthesized as a set of individual compounds that can give singly charge reference ions over the m/z range 140 to 772. Because these are individual compounds rather than a single compound that gives a distribution of oligomers or cluster ions, individual reference masses can be selected to bracket the mass of the analyte [11]. In conclusion, different calibrants are currently used, depending on the instrument manufacturer and/or specific application. The most common are polyethylene or polypropylene glycols, Ultramark 1621, phosphazines, mixtures of caffeine, a small peptide (MRFA) and myoglobin, or other mixtures of peptides and proteins.

13.3.4 APCI Source Tuning and Calibration

An atmospheric pressure chemical ionization (APCI) interface is generally considered extremely easy to optimize and operate. This is perhaps best proven by the fact that hardly any optimization of the interface parameters is reported

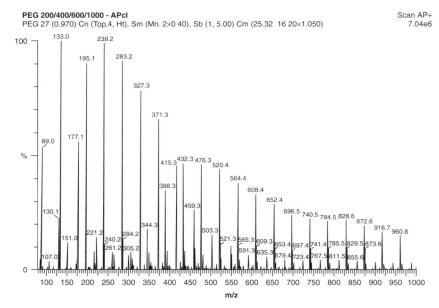

Figure 13.16. Calibration in APCI positive using PEG. (Courtesy of Micromass Quattro LC.)

in the many APCI application papers. The optimization for important interface parameters, such as the liquid flow rate, solvent composition, nebulizer and auxiliary gas flow, probe position, and vaporizer temperature, is less critical than, for instance, in ESI.

Reserpine is the most commonly used calibrant for APCI. The average molecular weight of reserpine is 608.7 and is generally injected using flow injection analysis (FIA) at a concentration of 10 pg/μL in 1% acetic acid in 50:50 methanol/water [6,7]. The use of PEG was described in the literature for calibration of an APCI-MS system in positive-ion mode (Figure 13.16) [5,10,25]. For calibration with PEG, it is best to use a large-volume injection loop (50 μL) with a solvent delivery system set up to deliver 0.2 mL/min of 50:50 acetonitrile/water or methanol/water through the injector and into the APCI source. An injection of 50 μL of PEG calibration solution lasts for approximately 15 s, allowing enough time to perform a slow scanning calibration. Since the PEG1000 has peaks from m/z 63 to 987, it is possible to calibrate over this mass range that is sufficient for the majority of applications with APCI. In many APCI operations, tuning strategies adapted from ESI are commonly used.

13.4   COMMON PROBLEMS AND SOLUTIONS

1. *LC-MS SOP and documentation.* The importance of having LC-MS SOPs and updated written records for LC-MS calibration and maintenance is particularly

critical during an audit [4]. Answering the following questions should help in verifying if the documentation related to LC-MS instrument is in full compliance with GLP and GMP regulations:

   a. Is the LC-MS within specification, and is the documentation to prove this available?
   b. If the LC-MS is not within specifications, by how much does it deviate?
   c. If the LC-MS is not within specifications, how long has this been the case?
   d. If the LC-MS is not within specifications, what action has been taken to overcome the defect?
   e. What standard has been used to test and calibrate the LC-MS before sample analysis?
   f. What action has been taken to guarantee the reliability of the data produced by the LC-MS used?

2. *Fit the purpose calibration.* It is common sense to check instrument performance each day, and GLP requirements simply formalize the performance and documentation of these checks. On the other hand, it is also important to use the right test (full calibration, verification, system suitability test, or instrument and method validation) to verify the performance and to avoid needlessly lengthy procedures. As already discussed (see Sections 13.2.3 and 13.3.1), it is not always necessary to perform a MS full calibration every day. For example, if a particular MS is used only to record complete full-scan mass spectra, a daily calibration or verification of the calibration of the m/z ratio scale is required. However, in the case where a MS is coupled with an LC and utilized primarily for the analysis of one or more analytes in the selected ion monitoring (SIM) mode, it does not always require a daily verification of the calibration. In this specific case it is quite common in LC-MS and LC-MS/MS applications to test only the following performance parameters: (a) sensitivity, (b) system precision, (c) linear dynamic range, (d) analytes retention time, and (e) chromatographic peak shape. All these parameters can be checked simultaneously by injections of system suitability solutions. Usually, if the system suitability test fails for nonchromatographic reasons, a possible deterioration of the m/z axis calibration should be taken into consideration and at least a verification of the calibration should be performed.

3. *Cross-contamination.* During the tuning and calibration procedures a tuning or calibration solution is infused using a syringe pump connected directly to the ion source or going into LC flow using a tee union. In general, the LC-MS is tuned and calibrated before data acquisition in either the ESI or the APCI mode. Due to the fact that many calibration solutions are quite "sticky," it is suggested to minimize the possibility of cross-contamination by using a different syringe and section of fused silica/peek tubing for the calibration solution and another for the tuning solution containing the analytes of interest. The infusion tuning/calibration procedures put a comparatively large amount of calibrants/analytes into the MS.

Therefore, before performing an analytical run to analyze the analyte, it is suggested that the LC-MS interface be cleaned. These problems are partially avoided if the FIA method (see Section 13.2.2) is used for tuning.

4. *Failed calibration.* There are number of reasons for a calibration to fail. If an automated calibration method is used, it is possible that the reference peaks are not recognized when the reference file and calibration file are compared. This can be due to the following reasons:

a. Degradation of the calibration solution.

b. Serious contamination of the calibration solution.

c. No flow of solvent into the source.

d. Multiplier set too low, so that the less intense peaks are not detected.

e. Reference solution running out, so that the less intense peaks are not detected.

f. Incorrect ionization mode selected. It should be checked that the data have been acquired with the right ion mode ESI positive or negative;

g. Intensity threshold set in the peak match parameters (see Section 13.3.1) too high. Peaks are present in the acquired calibration file but are ignored because they are below the threshold level.

h. Peak search range set in the peak match parameters (see Section 13.3.1) too small. The calibration peaks lie outside the limits set by these parameters;

i. Wrong reference file selected.

If the correct calibration parameters were used, and good calibration data were acquired, the instrument should be calibrated correctly. However, in some circumstances it is possible to meet the calibration criteria without matching the correct peaks. This situation is unusual, but it is always sensible to examine the on-screen calibration report to check that the correct peaks have been matched. These errors may occur when the following parameters are set:

a. The intensity threshold is set to 0.

b. The peak search range is too wide. A more intense peak than the reference could be included in the window.

c. If a contamination or background peak lies within one of the peak search windows and is more intense than the reference peak in that window, the wrong peak will be selected. Under some conditions this may happen more often when PEG and PPG are used as calibrants.

There are two ways to avoid this problem:

a. If the reference peak is closer to the center of the peak search range, this range can be narrowed until the contamination peak is excluded. It should be ensured that no other reference peaks are excluded.

b. If the reference peak is not closer to the center of the peak search range, or if by reducing the window other reference peaks are excluded, the calibration should be executed manually.

Before starting an automated calibration, as calibration solution infuses, the value of the ion current signal should always be checked to verify if the signal is present; or the signal is stable, varying by less than about 15% from scan to scan. If these conditions are not verified, the following troubleshooting measures should be tried:

a. If a fused-silica tube is used inside the ESI needle, it should be ensured that the fused-silica sample tube is in the position recommended by the operator's manual for that model or brand.
b. The interface should be inspected to ensure that the inlet is clean.
c. The ESI probe should be optimized for the flow rate used.
d. The solution entering the probe should be free of air bubbles.
e. The tubing and connectors should be checked for possible leaks.

REFERENCES

1. *Code of Federal Regulations*, Title 21, *Food and Drugs*, Office of the Federal Register, National Archives and Records Administration, Washington, DC, 1994, Part 58 (available from New Orders, P.O. Box 371954, Pittsburgh, PA, 15250-7954).
2. R. K. Boyd, J. D. Henion, M. Alexander, W. L. Budde, J. D. Gilbert, S. M. Musser, C. Palmer and E. K. Zurek, *J. Am. Soc. Mass Spectrom.*, **7**, 211–218, 1996.
3. *Guidelines for Writing an SOP for Mass Spectrometry*, prepared by the Measurements and Standards Committee of the American Society for Mass Spectrometry, Sept. 1, 1997 (available from *www.asms.org/mssop.h*).
4. L. Huber, *Good Laboratory Practice: A Primer for High Performance Liquid Chromatography, Capillary Electrophoresis, and UV–Visible Spectroscopy*, Hewlett-Packard Publ. 12-5091-6259E, 1993.
5. *User's Guide Micromass Quattro LC.*
6. *Operating the Sciex API3000: Operator Manual*, Apr. 1998.
7. *Operating the Sciex API4000: Operator Manual*, May 2001.
8. *Finnigan LCQ Deca: Operator Manual.*
9. M. A. Baldwin and G. J. Langley, *Org. Mass Spectrom.*, **22**, 561, 1987.
10. W. M. A. Niessen, *Liquid Chromatography–Mass Spectrometry*, Chromatographic Scientific Series, Vol. 79, Marcel Dekker, New York, 1999.
11. B. N. Pramanik, A. K. Ganguly and M. L. Gross, *Applied Electrospray Mass Spectrometry*, Practical Spectroscopy Series, Vol. 32, Marcel Dekker, New York, 2002.
12. K. L. Olson, K. L. Rinehart, Jr., and J. C. Cook, Jr., *Biomed. Mass Spectrom.*, **5**, 284–290, 1977.
13. L. Jiang and M. Moini, *J. Am. Soc. Mass Spectrom.*, **3**, 842–846, 1992.

14. M. Moini, *Rapid Commun. Mass Spectrom.*, **8**, 711–714, 1994.

15. R. Feng, Y. Konishi and A. W. Bell, *J. Am. Soc. Mass Spectrom.*, **2**, 387–401, 1991.

16. J. Zaia, R. S. Annan and K. Biemann, *Rapid Commun. Mass Spectrom.*, **6**, 32–36, 1992.

17. *CECA Hop. J. Mass Spectrom.*, **31**, 1314–1316, 1996.

18. D. J. Liberato, C. C. Fenselau, M. L. Vestal and A. L. Yergey, *Anal. Chem.*, **55**, 1741, 1983.

19. R. B. Cole and A. K. Harrata, *J. Am. Soc. Mass Spectrom.*, **4**, 546, 1993.

20. H. Fujiwara, R. C. Chott and R. G. Nadeau, *Rapid Commun. Mass Spectrom.*, **11**, 1547–1553, 1997.

21. A. P. Tinke, C. E. M. Heeremans, R. A. M. Van Der Hoeven, W. M. A. Niessen, J. Van Der Greef and N. M. M. Nibbering, *Rapid Commun. Mass Spectrom.*, **5**, 188, 1991.

22. F. Hsu, *Biol. Mass Spectrom.*, **21**, 363, 1992.

23. M. Moini, B. L. Jones, R. M. Rogers and L. Jiang, *J. Am. Soc. Mass Spectrom.*, **9**, 977–980, 1998.

24. S. Konig and H. M. Fales, *J. Am. Soc. Mass Spectrom.*, **10**, 273–276, 1999.

25. H. Y. Lin, G. J. Gonyea and S. K. Chowdhury, *J. Mass Spectrom.*, **30**, 381, 1995.

26. J. F. Anacleto, S. Pleasance and R. K. Boyd, *Org. Mass Spectrom.*, **27**, 660, 1992.

27. R. B. Cody, J. Tamura and B. D. Musselman, *Anal. Chem.*, **64**, 1561, 1992.

14

KARL FISHER APPARATUS AND ITS PERFORMANCE VERIFICATION

Rick Jairam, Robert Metcalfe, Ph.D., and Yu-Hong Tse, Ph.D.
GlaxoSmithKline Canada, Inc.

14.1 INTRODUCTION

The Karl Fisher titration is one of the most common and most sensitive methods used in the analytical laboratory. The titrimetric determination of water is based on the quantitative reaction of water with an anhydrous solution of sulfur dioxide and iodine in the presence of a buffer that reacts with hydrogen ions. This titration is a two-stage process:

$$SO_2 + MeOH + RN \rightarrow (RNH)SO_3Me \qquad (14.1)$$

$$(RNH)SO_3Me + I_2 + H_2O + 2RN \rightarrow (RNH)SO_4Me + 2(RNH)I \quad (14.2)$$

where RN is a base, typically pyridine or imidazole. Reaction (14.1) reaches equilibrium and produces methylsulfite as the reaction intermediate. Reaction (14.2) the redox process, is very rapid. From equation (14.2) the direct relation between water and iodine consumption can be seen, which enables the amount of water to be determined. Complete esterification of the sulfur dioxide with the alcohol, and the ability of the base to neutralize the methyl sulfurous acid, are the key requirements for the reaction above to be stoichiometric.

Analytical Method Validation and Instrument Performance Verification, Edited by Chung Chow Chan, Herman Lam, Y. C. Lee, and Xue-Ming Zhang
ISBN 0-471-25953-5 Copyright © 2004 John Wiley & Sons, Inc.

Pyridine was used in the beginning of the development of the method. The reaction was slow and the endpoint unstable because of weak basicity of pyridine. The pyridine system buffers at about pH 4. A stronger base, imidazole, has been used to replace pyridine since it gives a faster response and has the advantages of lower toxicity and decreased odor. The optimal pH range for the SO_2 imidazole buffer is at pH 6. It is important that the pH of the Karl Fisher reaction be maintained within the range 5 to 7. Outside this recommended pH range, the endpoint may not be reached.

There are two types of Karl Fisher titrations: volumetric and coulometric. *Volumetric titration* is used to determine relatively large amounts of water (1 to 100 μg) and can be performed using the single- or two-component system. Most commercially available titrators make use of the one-component titrant, which can be purchased in two strengths; 2 mg of water per milliliter of titrant and the 5 mg of water per milliliter of titrant. The choice of concentration is dependent on the amount of water in the sample and any sample size limitations. In both cases, the sample is typically dissolved in a methanol solution. The iodine/SO_2/pyridine (imidazole) required for the reaction is titrated into the sample solution either manually or automatically. The reaction endpoint is generally detected bivoltametrically.

Coulometric titration is used to determine relatively low concentrations of water (10 μg to 10 mg) and requires two reagents: a catholyte and an anolyte (the generating solution). The iodine required for the reaction is generated in situ by the anodic oxidation of iodide.

$$2I^- \rightarrow I_2 + 2e^- \tag{14.3}$$

The iodine then reacts with the water that is present. The amount of water titrated is proportional to the total current (according to Faraday's law) used in generating the iodine necessary to react with the water. One mole of iodine reacts quantitatively with 1 mol of water. As a result, 1 mg of water is equivalent to 10.71 C. Based on this principle, the water content of the sample can be determined by the quantity of current that flows during the electrolysis. For this reason, the coulometric method is considered an absolute technique, and no standardization of the reagents is required.

14.2 SCOPE OF CHAPTER

The Karl Fisher instrumentation and its performance verification are discussed in this chapter. The instrumentation, calibration practices, and common difficulties that are encountered are presented. Neither method validation nor method specific problems are discussed.

14.3 INSTRUMENTATION

Karl Fisher apparatus has to be designed to exclude moisture, deliver titrant, and to detect the endpoint. The air in the system is kept dry with a suitable desiccant, and the titration vessel may be purged by means of a stream of dry nitrogen or air. For endpoint detection, most commercially available units use a bivoltametric method to indicate that the endpoint has been reached. For this method a constant current of about 20 μA is applied across a pair of platinum electrodes that are about 2.5 mm apart. As the titration proceeds, water reacts with iodine and is consumed. When the endpoint is reached and all the water is consumed, there is a buildup of free iodine in solution. The free iodine causes ionic conduction in the solution. As a result, the voltage must be reduced to keep the polarization current constant [1,2]. When the voltage drops below a defined value, the titration is stopped. Figure 14.1 shows a schematic diagram of a typical KF titrator.

In some cases (see below) a KF drying oven is required to get the water from a sample into the titration vessel. For these special cases, a solid sample (usually) is placed into a specially designed KF oven where the sample is heated, and the water goes into the vapor phase. A stream of dry carrier gas (usually, N_2 or air) sweeps the liberated moisture into the reaction vessel, where it is titrated by either the coulometric or the volumetric method. It is critical that the carrier gas is dry and that there are no leaks along the pathway to the reaction vessel. Passing the carrier gas over activated molecular sieve prior to the sample will ensure that the gas is dry.

The KF reaction depends on free water available for titration. Instruments may be designed to homogenize the sample to release water prior to the titration. In

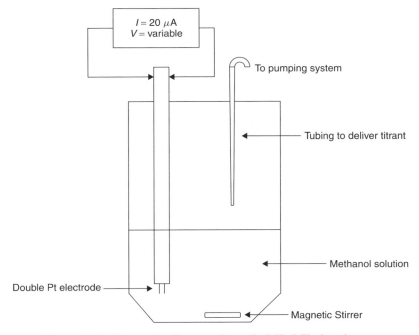

Figure 14.1. Schematic diagram of a typical Karl Fischer titrator.

this case, a high-speed blender may be used directly in the titration vessel. This is a particularly useful accessory for determination of water in tablets or certain food products that are difficult to prepare for analysis.

14.3.1 Performance Verification

The performance verification of Karl Fisher apparatus should include checks for the accuracy and precision of the instrument. The linearity of the instrument should be determined at installation. The first step is to standardize the instrument (see Section 14.3.2); pure water is sufficient for this purpose. Sodium tartrate dihydrate standard (water content 15.66 ± 0.05%) can be used to assess the accuracy, precision, and linearity of the instrument. Typically, one would measure the water content of at least five samples, over the intended instrument user range. For example, the water content of sodium tartrate dihydrate samples that were 65 mg (ca. 10 mg H_2O), 195 mg (ca. 30 mg H_2O), 325 mg (ca. 50 mg H_2O), 455 mg (ca. 70 mg H_2O), and 650 mg (100 mg H_2O) could be determined. Calculate the percent water to assess the instrument's accuracy. The results should be within 98 to 102% of 15.66% water. Determine the % RSD of the percent water found to assess the precision of the instrument. The % RSD should be less than or equal to 1%. Finally, plot the expected water content versus the percent water content to assess the linearity of the instrument's response. A correlation coefficient (r) value of 0.999 or greater is acceptable.

14.3.2 Standardization

Standardization should be performed daily, since the titrant will absorb moisture over time. The standardization is typically performed using one of three standards: (1) disodium tartrate dihydrate (15.66% water, w/w), (2) commercially available water standards with a certified concentration of water (typically, 10 mg of water per gram of standard), or (3) pure water. In all of these cases, a suitable balance should be used, preferably one that gives at least ±0.1 mg of resolution.

When disodium tartrate dihydrate is used, samples ranging from 50 to 120 mg are dissolved in methanol, and the concentration of the KF reagent is determined. Based on volume of titrant used, the weight of the sample and the percent water in the disodium tartrate dihydrate (15.66% w/w), the standardization factor can be calculated. One pitfall with this method is the solubility of the disodium tartrate dihydrate in methanol. It is recommended that the disodium tartrate dihydrate be finely divided and that a suitable extraction time be given for the solids to dissolve.

If a water standard is used, samples of the water standard ranging from 1.0 to 1.5 g are injected into the methanol via syringe. This method is best performed by opening the ampoule that contains the water standard, rinsing out a 10-mL syringe with about 1 mL of the standard, and then filling the syringe with the remaining contents of the ampoule. Aliquots of approximately 1 mL are then injected into the vessel, and the true weights are determined by difference.

Standardization of the KF titrant using purified water is widely accepted and probably the most commonly used in analytical laboratories. Using a suitable glass syringe, a known volume of purified water can be added in the titration vessel, and the KF value for the titrant can be determined. This method is highly operator dependent and is not recommended for measurements where the accuracy of the determination is critical. The preferred method would be to add a known quantity of water by weight.

In the coulometric method, standardization is not necessary, since the current consumed can be measured absolutely. However, a standard with known water content should be checked periodically to ensure that the system is functioning properly. In this case, a certified water standard is generally used, and the amount of water is determined and compared with the amount that is certified to be present. Some coulometric titrators are equipped with an oven for liberating the moisture from samples that are either insoluble in methanol or that react with I_2, methanol, or one of the other reagents. Solid standards (e.g., potassium citrate monohydrate) are available for checking the oven, and this check is performed after the coulometer function has been verified.

14.3.3 Measurement of Samples

The handling of the sample, its storage, and the amount of sample used are very important to consider when making a KF measurement. When taking a sample from the bulk material, great care must be taken to exclude atmospheric moisture, which is one of the most common sources of error. If the sample absorbs or desorbs water during its handling, the true moisture content can no longer be determined. The sample must also contain the average amount of water that is contained in the material as a whole. This is a particular problem when dealing with nonpolar liquids such as fats or oils where the water is floating on the top or settled to the bottom of the container. In this situation the liquids must be mixed thoroughly just prior to sampling to disperse the water evenly.

Once a sample is taken, it is best to determine the water as quickly as possible. If a sample must be stored, it is best to keep it in a small, tightly sealed glass bottle with a screw cap. Glass is better than plastic since plastic is not water vapor–tight. The size of the bottle should be as small as possible to minimize the gas space that is above the sample. A smaller gas space will mean less moisture vapor.

The size of the sample to be used depends on the expected water content and the degree of accuracy that is desired. If a high level of accuracy is required, a large sample size should be used to minimize the effect of atmospheric moisture. For titrimetric KF, the *United States Pharmacopoeia* recommends using a sample that will contain between 10 and 250 mg of water. The minimum amount of water is generally agreed to be 10 mg.

14.4 COMMON PROBLEMS AND SOLUTIONS

Atmospheric moisture is a major cause of error in Karl Fischer titrations. Moisture can enter the sample, titrant, and titration vessel. The apparatus and titrant must be sealed against atmospheric moisture and situated away from high-humidity areas of the laboratory. Desiccants and dry purge gases can be used to reduce moisture ingress. Proper conditioning of the apparatus is required to remove moisture present in the apparatus before the titration. Moisture will, however, enter the apparatus during introduction of the sample in a normal lab environment. With high-speed blending, this introduction of moisture can cause bias due to homogenizing air containing water into the sample. To reduce bias when high-speed blending is used, it is necessary to apply background correction or purge the sample compartment to exclude moisture.

The titrant delivery system can also be a source of error. The delivery system needs to be precise in the number of steps required to titrate a fixed amount of water. The delivery of titrant is typically controlled by either a peristaltic pump or an automatic buret. Both are acceptable methods for the delivery of titrant; however, the peristaltic pump relies on the contact between the pump rollers and the tygon tubing, which is pressed into position by a platen. Therefore, the quality of the tubing that is used is critical to the accuracy of the instrument, so this tubing should be examined frequently to make sure that it is in good working order (free of blockages and not pinched to the point that flow is restricted). The pump rollers should also be examined to make sure that they are operating well, and the platen should be examined to make sure that it holds the tubing tightly against the rollers. The automatic buret is actually a syringe driven by a metered servomotor. For this type of delivery system the key concerns are that the servomotor is operating properly (i.e., delivering the correct amount of liquid) and that there are no leaks, blockages, or air bubbles in the lines or in the syringe itself.

Depending on the samples being titrated, electrode contamination may be an issue. The electrodes may become coated when substances such as oils and sugars are titrated. This results in delayed endpoint detection and an overtitration. A dark brown color on the electrodes indicates that they should be cleaned. The coatings can usually be removed by polar organic solvents or by cleaning the platinum physically.

Side reactions can cause major problems in the determination water [3]. Some chemical species react with methanol to produce water that will lead to higher water results, some react with water (low results), and some react with iodine (high result). Aldehydes and ketones react with methanol to produce acetals and ketals. Water is a by-product of this reaction, and its production leads to erroneously high water contents:

Acetal formation: $CH_3CHO + 2CH_3OH \rightarrow CH_3CH(OCH_3)_2 + H_2O$ (14.4)

Ketal formation : $(CH_3)_2CO + 2CH_3OH \rightarrow (CH_3)_2C(OCH_3)_2 + H_2O$ (14.5)

This can be remedied by using commercially available one-component reagents that contain 2-methoxyethanol and 2-chloroethanol rather than methanol. In the presence of SO_2 and base, aldehydes, undergo what is called the *bisulfite addition*.

This reaction consumes water.

Bisulfite addition: $CH_3CHO + H_2O + SO_2 + NR \rightarrow HC(OH)SO_3(HNR)$

(14.6)

The kinetics of this type of reaction are slow, so this can be avoided by starting the reaction immediately after sample addition.

Many samples have redox potentials such that they can be oxidized by iodine. Therefore, the iodine in the titrant may be consumed by readily oxidizable samples that will give a false high value for the water content. Some common substances that can be oxidized by iodine are ascorbic acid, arsenite (AsO_2^-), arsenate (AsO_4^{3-}), boric acid, tetraborate ($B_4O_7^{2-}$), carbonate (CO_3^{2-}), disulfite ($S_2O_5^{2-}$), iron(II) salts, hydrazine derivatives, hydroxides (OH^-), bicarbonates (HCO_3^-), copper(I) salts, mercaptans (RSH), nitrite (NO_2^-), some metal oxides, peroxides, selenite (SeO_3^{2-}), silanols (R_3SiOH), sulfite (SO_3^{2-}), tellurite (TeO_3^{2-}), thiosulfate ($S_2O_3^{2-}$), and tin(II) salts. For situations such as these where the material under analysis reacts with iodine, an oven can be used to liberate the moisture from the sample, which is then carried into the reaction vessel and titrated without interference.

REFERENCES

1. *United States Pharmacopoeia*, USP 25, Chapter ⟨921⟩, Water Determination.
2. *Fundamentals of the Volumetric Karl Fischer Titration*, Mettler Toledo Application Brochure 26.
3. *Riedel-deHaen Hydranal Manual.*

15

THE pH METER AND ITS PERFORMANCE VERIFICATION

Yu-Hong Tse, Ph.D., Rick Jairam, and Robert Metcalfe, Ph.D.
GlaxoSmithKline Canada, Inc.

15.1 INTRODUCTION

The measurement of pH is one of the most common tests performed in a chemical laboratory since many chemical processes and properties are pH dependent. Examples of these processes are the kinetics of chemical reactions, the spectrum of certain dyes, as well as the solubility and/or bioavailability of many chemicals.

The definition of pH represents the measure of the activity of hydrogen ions in a solution at a given temperature. It is derived from a combination of "p" for the word *power* and "H" for the symbol for the element hydrogen. Mathematically, pH is the negative log of the activity of hydrogen ions. This relationship is illustrated in the formula

$$pH = -\log[aH^+] \tag{15.1}$$

The term *activity*, aH^+, is used because pH reflects the amount of available hydrogen ions instead of the concentration of hydrogen ions. In an aqueous solution, the following equilibrium exists between hydrogen ions (H^+) and hydroxide ions (OH^-):

$$H_2O \leftrightarrow H^+ + OH^- \quad \text{or} \quad 2H_2O \leftrightarrow H_3O^+ + OH^- \tag{15.2}$$

The greater the number of hydrogen ions (H^+), the more acidic the solution. The greater the number of hydroxide ions (OH^-), the more alkaline or basic the

Analytical Method Validation and Instrument Performance Verification, Edited by Chung Chow Chan, Herman Lam, Y. C. Lee, and Xue-Ming Zhang
ISBN 0-471-25953-5 Copyright © 2004 John Wiley & Sons, Inc.

solution. The pH scale for aqueous solutions ranges from 0 to 14 pH units, with pH 7 being neutral. Purified water typically has a pH value of about 7.

The only accurate way to measure pH is with potentiometric electrodes attached to a voltmeter (pH meter). Potentiometric electrodes monitor the changes in voltage caused by differing concentrations of hydrogen ions [H^+]. These electrodes consist of a pH-sensing electrode (most commonly a glass electrode) and a reference electrode (Figure 15.1). Ideally, the pH-sensing electrode will only be sensitive to hydrogen ions. The reference electrode provides a constant potential independent of the concentration of hydrogen ions.

The following is a schematic of an electrochemical cell consisting of an AgCl-coated Ag wire reference electrode and a pH-sensing glass membrane electrode. The bars represent phase boundaries.

$$Ag|AgCl|4 \ M \ KCl|test \ solution|glass \ membrane|pH \ 7 \ buffer,$$

$$4 \ M \ KCl|AgCl|Ag \qquad (15.3)$$

The potential (E) generated by the glass electrode can be calculated theoretically using the Nernst equation,

$$E = E^0 + (2.303RT/nF)(\log[aH^+]) \qquad (15.4)$$

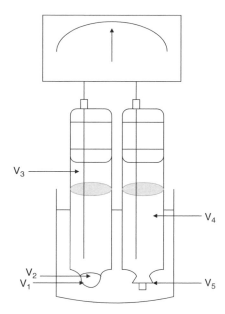

Figure 15.1. Potentiometric measurement for pH. V_1, glass membrane; V_2, inner buffer solution; V_3, internal reference electrode relative to internal buffer; V_4, external reference electrode; V_5, diaphragm.

where E is the measured electrode potential from the glass electrode, E^0 is related to the potential of the reference electrode, $(2.303RT/nF)$ is the Nernst factor and $-\log[aH^+]$ is the pH. The Nernst factor, $2.303RT/nF$, includes the gas law constant (R), Faraday's constant (F), the temperature in kelvin (T), and the number of electrons transferred in the electrochemical process (n). For pH, where $n = 1$, the Nernst factor is $2.303RT/F$. Since R and F are constants, the Nernst factor and the electrode behavior (the slope) are temperature dependent. At $25°C$ ($298°K$), equation (15.4) can be rewritten as follows:

$$E = E^0 + (\text{slope} \times \text{pH}) \tag{15.5}$$

where the slope has the value of -59.16 mV/pH unit.

Accurate measurement of pH is critically dependent on good analytical procedures, a fact that may not be appreciated by laboratory personnel [1,2]. The assumption is often made that if the electrode has been calibrated, there will be no variability in pH between laboratories. The pH measurement can erroneously be seen as merely dipping the electrode into the analyte and recording the value. In 1985, Davidson and Gardner [3] drew the following conclusion from their study "Interlaboratory Comparisons of the Determination of pH in Poorly Buffered Fresh Waters":

> Considerable bias errors may be due to the preparation of standards, as well as inaccuracies associated with the actual measurements. Undoubtedly, care in the selection and initial testing of electrodes will improve the quality of results. Most important, however, is the unambiguous description of preparation and measurement procedures, and the adoption of routine analytical quality control. The quality of pH data, like those from any other analytical determination, will be greatly improved by strict adherence to a rigorously defined proven routine. Although implementing a program of quality control will decrease errors, it will not ensure the accuracy of the determination.

Therefore, analysts must remember that reliable and accurate pH measurements depend on the quality of the equipment used, the maintenance of the electrodes, the accuracy of the calibration, and the adherence to proper procedures.

15.2 SCOPE OF CHAPTER

In this chapter, a brief introduction to the pH meter and the performance verification parameters is presented. Some good laboratory practices and method development are discussed. The solution to commonly encountered problems is also offered. It is not within the scope of this chapter to discuss each factor in detail but rather, to draw attention to and raise the reader's awareness of the potential problems in pH measurement.

15.3 INSTRUMENTATION

15.3.1 Electrodes

The measurement of pH is carried out using a sensing electrode, which is sensitive to hydrogen-ion activity and a reference electrode. Combination electrodes incorporating both of these electrodes are also suitable for most applications. Separate reference and sensing electrodes are normally used only for high-precision research applications.

Sensing Electrode. Ideally, a hydrogen electrode should be used for pH measurements. The hydrogen electrode consists of a platinum electrode across which hydrogen gas is bubbled at a pressure of 101 kPa. Due to the difficulties associated with the use of the hydrogen electrode and the potential for electrode fouling, glass electrodes are the most commonly used sensing electrode. The integral components of a glass electrode are an electrode membrane that responds to pH, an internal electrode (usually AgCl-coated Ag wire), and an internal conductive solution (usually, potassium chloride solution) maintained at pH 7.

The most critical item in this system is the electrode membrane: (1) the membrane glass must generate a potential that responds accurately to the pH of the solution; (2) the membrane must not only be responsive to acidity and alkalinity, it must also be resistant to their chemical attack; and (3) the electrical resistance of the membrane must not be too great. The glass membrane is available in different sizes, shapes, and glass composition. Generally, the glass electrode contains by weight 72.2% SiO_2, 6.4% CaO, and 21.4% Na_2O [4]. Lithium is sometimes used to replace sodium to increase the operational range of the electrode. The composition of glass dictates the response time, selectivity, and sensitivity of the electrode.

The glass electrode is resistant to interference from colloidal matter, oxidants, reductants, and high salinity, but some common ions, such as sodium, lithium, and potassium, do interfere at high pH (>11). Disadvantages of the glass electrodes are that routine maintenance is required (see the discussion below) and they are fragile. Electrodes other than glass and hydrogen are known but are not practical to use [4].

To function properly, a glass membrane requires hydration. Upon hydration of the glass membrane on the glass electrode, a gel layer will be formed on the interface between the glass electrode and the solution (Figure 15.2). Hydrogen ions in the solution can exchange with the sodium in the gel layer. The inner membrane layer interacts with the internal electrode solution, which has a constant hydrogen ion activity, and hence potential, while the outside layer will interact with the test solution, in which the hydrogen ion activity (pH) can vary. It is this potential difference across the membrane that determines the pH.

Reference Electrode. Any potentiometric measurement requires a reference electrode to complete the electrical circuit. A piece of bare wire is not recommended because the voltage of its surface could change in an unpredictable fashion with

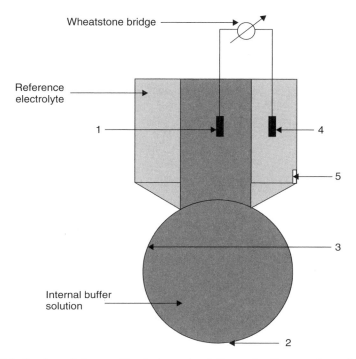

Figure 15.2. Design of the combined electrode. 1, Internal reference electrode, usually Ag|AgCl; 2, outer glass membrane; 3, inner glass membrane; 4, external reference electrode, usually Ag|AgCl; 5, diaphragm.

time and a test sample. A good reference electrode should provide a defined stable reference potential point and be independent of the analyte solution. A reference electrode normally consists of a reference element, which is stored in a defined electrolyte. This electrolyte must be in contact (liquid junction) with the analyte solution. Depending on the physical construction of the liquid junction, the composition of the electrolyte, and the reference element, different reference electrodes are available (Table 15.1). A porous pin forms the most common type of liquid junction. Circular ceramic junctions, sleeve junctions, or an open junction through a thin glass tube are also available. These will ensure a higher outflow of salt bridge solution, which is beneficial when measuring in solutions of very high or very low ionic strength. Some common reference electrodes are $Hg|Hg_2Cl_2$, $Ag|AgCl$, $Hg|Hg_2SO_4$, $Hg|Hg_2Cl$, but the $Ag|AgCl$ electrode is the most commonly used because of safety issues associated with the use of mercury.

Combination Electrode. Combining the sensing and reference electrode as one single electrode provides convenience during routine testing. The rugged single-body construction is beneficial in routine analysis and helps in measurement of the pH of small sample volumes. It also eliminates the potential problem of a

Table 15.1. Advantages and Disadvantages of Some Common Reference Electrodes

Electrode Type	Advantage	Disadvantage
Reference electrode with KCl salt bridge	Good for aqueous samples, low junction potential	Not good for samples that are sensitive to K^+ and Cl^- ions; use is limited to room temperature
Reference electrode with open liquid junction	Advantages are similar to a KCl salt bridge, but it is also good for highly viscous samples	Not good for samples that are sensitive to K^+ and Cl^- ions; use limited to room temperature
Reference electrode with LiCl salt bridge	Good for nonaqueous sample, low junction potential	Not good for samples that are sensitive to K^+ and Cl^- ions; use limited to room temperature
Ag/AgCl reference electrode	Good for high-temperature conditions above 60°C	Not good for sample that is sensitive to K^+ and Cl^-
Reference electrode with K_2SO_4 salt bridge solution	Good for low-viscosity samples that are sensitive to K^+ and Cl^- sample	
Double junction reference electrode with sleeve	Good for viscous samples that are sensitive to K^+	

temperature gradient between the sensing and reference electrodes. However, all the comments discussed above on individual electrodes are also applicable to the electrode combination.

15.3.2 pH Meter

The pH meter is a specialized voltmeter that measures the potential difference (in mV) between the sensing and reference electrode and converts it to a display of pH. To provide an accurate measurement of the voltage of an extremely high resistance electrode (10^8 Ω) [5], this specialized voltmeter must be designed with high input resistance or impedance characteristics (100 times that of the electrode used). Since the measurement potential difference per pH change is very small (59.16 mV/pH unit at 25°C), a reliable amplifier in the pH meter is also essential. It should be sufficiently sensitive to detect changes of at least 0.05 pH unit (or 3 mV).

15.3.3 Buffer

The use of buffers for the calibration of a pH electrode is a critical step in any pH measurement. Buffer solutions of different pH values can be prepared or are commercially available. They should be traceable to national standards with ±0.01

pH unit of accuracy (NIST, BS). High-precision buffers can be prepared from high-quality reagents (ACS, AnalaR) using recognized standard methods (BP/EP, USP, AOAC), but it is important to use them within two months. These solutions may be stored in tight, chemically resistant containers, such as type I glass bottles. Protection from absorption of carbon dioxide is critical for the shelf life of alkaline buffers. Most commercially available standards contain preservatives to prevent the growth of microbes.

15.3.4 Temperature Effect

The most common cause of error in a pH measurement is temperature. Most pH meters contain an automatic temperature compensation (ATC) function to adjust the pH response according to the Nernst factor when the temperature of the sample differs from the calibration temperature. If the temperatures differ by more than 10°C, there will be an error of 0.15 pH unit or more (between pH 3 and 11) without the ATC function.

When the temperature of the testing solution is different from that of the electrode, the temperature of the electrode will drift until they are equal. The speed of this change depends on the physical design of the electrode and the heat transferring ability of the fill solution inside the electrode. As indicated in equation (15.4), the slope of the electrode is temperature dependent. A change in temperature from 50°C to 100°C will result in a slope change of about 10 mV/pH unit (Figure 15.3).

The pH of the standard buffer and sample can vary with temperature due to shifts in their chemical equilibrium. Since the actual pH of the most common

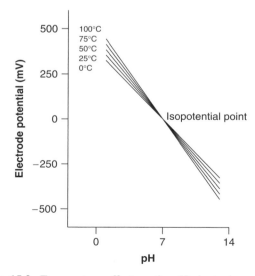

Figure 15.3. Temperature effect on the pH electrode response.

standard buffer solutions is known, this problem can easily be resolved. The effect of the temperature on the pH of a known sample should be part of the development of a pH measurement method. Although most pH meters are equipped with an ATC function to compensate for temperature differences between the sample and standard buffer, a temperature probe can also be the source of a potential problem for pH measurements if there are different response factors for the temperature probe and the pH electrode. Furthermore, the temperature may be inhomogeneous inside the testing solution.

15.4 PERFORMANCE VERIFICATION AND PARAMETERS FOR CALIBRATION

Although the pH meter (electrode) is one of the simplest instruments in the laboratory and its performance verification may be as simple as following the calibration procedures, failure to follow these procedures can result in significant error in the pH measurement.

15.4.1 General pH Calibration Procedures

Electrode. Your pH meter user's guide is a valuable source of information on how to calibrate the pH electrode, since specific instructions or steps may be required to obtain reliable pH data. A two-point or multipoint calibration, using buffers is required, before the pH is measured. Daily calibration before use is necessary to determine the slope of the electrode. This serves the dual purpose of determining if the electrode is working properly and storing the slope value in the instrument's memory. A calibration of at least two points is needed to determine the slope of the electrode.

The standard procedure for a two-point calibration of a pH electrode is to select two buffers that bracket the expected sample pH value, whose difference in pH does not exceed 4 units. One of these buffers must be at pH 7. The pH of the second buffer solution should be within ± 0.07 pH unit of the manufacturer's stated value. If a larger deviation is noted, examine the electrodes and replace them if they are faulty. Adjust the slope or temperature control to make the pH value observed identical with the value stated. Standardization should be repeated until both buffer solutions for standardization give observed pH values within 0.02 pH unit of the stated value without further adjustment of the controls.

Most pH meters will display the electrode response (slope) as a percentage of the theoretical value (59.16 mV/pH unit at 25°C). The electrode response should not be less than 95% or more than 105% of the theoretical value at a given buffer temperature. Contamination of the electrode or changes in the liquid junction potential will typically lower the electrode response to below 95%. Some common procedures to regenerate the electrode are listed below. Replacement of the electrode will be required if regeneration is not successful. If there are more than 4 pH units between the two calibration buffers used, immerse the electrodes in a third buffer of intermediate pH. The pH measured should be within ±0.05 pH unit of the pH of the buffer, corrected for temperature.

E^0 [equation (15.4)] is also referred to as the *offset*, the *zero potential point*, or the *isopotential point*, since theoretically it is defined as the pH that has no temperature dependence. Most pH electrode manufacturers design their isopotential point to be 0 mV at pH 7 to correspond with the temperature software in most pH meters. The offset potential is often displayed after calibration as an indication of electrode performance. Typical readings should be about 0 ± 30 mV in a pH 7 buffer. In reality, E^0 is composed of several single potentials, each of which has a slight temperature coefficient. These potentials are sources of error in temperature compensation algorithms.

The following detailed procedure for measuring pH may sound simple, but failure to follow the steps properly is a key factor for poor precision.

- *Step 1:* Remove the electrode from its storage solution and rinse with purified water. Dab the electrode with tissue paper to remove excess water that could dilute the solution to be tested. *Do not wipe the glass bulb.* Place the electrode (and the temperature probe if applicable) into the first buffer to be tested. Wait for a stable response from the instrument. (*Note:* If using a refillable electrode, open the fill-hole cover during calibration or any measurement to allow a uniform flow of electrode filling solution. Close the fill hole when the electrode is not in use.)
- *Step 2:* Calibrate the meter to read the temperature-corrected value of the first buffer. (*Note:* The filling solution level must be higher than the sample level to maintain the flow of filling solution. At least 1 inch above sample height is recommended.)
- *Step 3:* Rinse the electrode and probe and dab them with a tissue to remove any excess water. An alternative rinsing procedure is to rinse with the next solution to be tested. Place the electrode and temperature probes in the next buffer. Wait for the reading.
- *Step 4:* Calibrate the meter to read the temperature-corrected value of the second buffer.
- *Step 5:* Rinse the electrode(s) and temperature probe as before. Place the electrode and temperature probe in the sample and wait for a stable reading.

An electrode verification procedure should be performed as frequently as possible since electrodes are known to drift with time. Recalibrate if the electrode response has drifted outside the value specified (see the limit cited above).

pH Meter. The pH meter should be calibrated annually, preferably more frequently. Since a pH meter is a special type of voltmeter, calibration can be carried out using traceable potential simulators in the normal operation of the pH range (±300 mV), but these are rather expensive. The following performance check procedures are usually performed to identify any potential problems with the pH meter.

- *Step 1:* Change the pH meter to the mV mode. Short the circuit by connecting the inputs (sensing and reference) together. The pH meter should display only a few mV: ideally, 0.0 mV. Record the reading.
- *Step 2:* Repeat step 1 by connecting a resistance of 1 GΩ between the inputs. The reading should not differ by more than about 1 mV from that in step 1.
- *Step 3:* Repeat step 1 by connecting the 1.5-V dry cell between the electrode inputs. Note the reading on the display.
- *Step 4:* Repeat step 3 with connecting a resistance of 1 GΩ with the 1.5-V dry cell in series. The reading should not differ by more than a few mV that in step 3.

15.4.2 Developing a pH Measurement Procedure

A systemic approach should be used in the development of a precise and accurate method for pH determination. Consideration should be given, but not limited to, the following: the specification of the pH meter, the suitability of the electrode for the measurement, the requirement for temperature compensation, the buffer solutions, the reagents (purified water, electrolyte solution, and cleaning solutions), glassware cleanliness, and homogeneity of the solution.

Choosing the Right Electrode. Choosing the correct dimensions of the electrodes for your experiment is straightforward since it will dependent on your sample size and sample container. However, the type of electrode used in your method will depend on the matrix of your sample and the measurement conditions for the sample preparation. Ideally, sample solutions will have a pH in the range 2 to 12, a temperature between 10 and 50°C, and an ion concentration between 0.5 and 4 M. Most common pH electrodes can provide reliable and accurate measurements under these conditions. Measuring the pH outside the range 2 to 12 requires special electrodes.

Poorly Conducting Samples. Samples with low ion concentrations (less than a few mM) will result in poor conductivity of the sample solution. Low conductivity not only causes a slow response time from the electrode but also creates a diffusion potential between the reference electrolyte and the measuring solution which results in an inaccurate pH reading. Using circular ground junctions that create optimal contact between the reference electrolyte and measuring solution can eliminate the problem. Adding conductive-free ions such as a few drops of saturated KCl solution into the sample solution will also improve its conductivity. Increasing the temperature can increase the conductivity of the solution. Low-ionic-strength buffers should also be used to decrease errors during calibration. The drifts of the electrode should also be monitored during measurement. A change of less than 0.05 unit/min should be acceptable, and successive measurements should agree within 0.1 unit.

Semiaqueous or Nonaqueous Solutions. Although the measurement of pH in mixed solvents (e.g., water/organic solvent) is not recommended, for a solution containing more than 5% water, the classical definition of a pH measurement may still apply. In nonaqueous solution, only relative pH values can be obtained. Measurements taken in nonaqueous or partly aqueous solutions require the electrode to be frequently rehydrated (i.e., soaked in water or an acidic buffer). Between measurements and after use with a nonaqueous solvent (which is immiscible with water), the electrode should first be rinsed with a solvent, which is miscible with water as well as the analyte solvent, then rinsed with water. Another potential problem with this type of medium is the risk of precipitation of the KCl electrolyte in the junction between the reference electrode and the measuring solution. To minimize this problem, the reference electrolyte and the sample solution should be matched for mobility and solubility. For example, LiCl in ethanol or LiCl in acetic acid are often used as the reference electrode electrolyte for nonaqueous measurements.

Protein-Rich Solution. When analyzing a sample with a high protein concentration, interference between the protein and the reference electrolyte may be a problem. Use of a suitable electrolyte is critical for accurate measurement.

Semisolid (e.g., Ointment or Syrup). The diaphragm (see Figure 15.1) is subject to contamination with semisolid substances and must be cleaned properly after measurement. When analyzing an ointment sample, the electrode should be cleaned with ethanol, rinsed with purified water, and the diaphragm rinsed with electrolyte after each measurement. The electrode should not be left in samples with a high sugar content for any length of time since dehydration of the gel layer could result. Rinsing the electrode well and soaking the electrode between measurements will increase the precision of the measurements.

Interference. Most common reference electrodes contain K^+ and Cl^- ions, which provide good ionic transport within the junctions. Other reference electrodes are also available for those samples that are sensitive to these ions (i.e., Hg_2SO_4 with a K_2SO_4 salt bridge, or reference electrodes with a double salt bridge construction).

Additional Sample Measurement Requirements. The sample measurement should ideally be made under the same conditions as the electrode calibration.

The difference in the conductivity of the calibration buffers and sample can cause a very large error on the sample measurement, due to junction potentials in different environments. Solid samples should be dissolved in purified water. It is necessary that the water be carbon dioxide-free. The presence of dissolved carbon dioxide will cause significant bias in the measurement of samples with low buffering capacity. For pH measurements with an accuracy of 0.01 to 0.1 pH unit, the limiting factor is often the electrochemical system (i.e., the characteristics of the electrodes and the solution in which they are immersed).

15.5 SOME GOOD LABORATORY PRACTICES

A regular maintenance schedule and proper storage of the pH electrode ensures proper performance, helps extend the life of the electrode, and avoids the cost of replacements. On a weekly basis the electrodes should be inspected for scratches, cracks, deposits, and membrane–junction deposits. The reference chamber of refillable electrodes should be drained, flushed with fresh filling solution, and refilled. (*Note*: The chamber should only be filled to about 75 to 80% of capacity, to allow expansion during temperature change.) This maintenance procedure will keep the electrode ready to use and improve the lifetime of the electrode. A standard glass electrode in normal use can last for up to two years.

15.5.1 Cleaning

Electrodes should be cleaned according to the manufacturer's instructions. The following procedures may be used in the absence of specific instructions. Soaking the electrode in 0.1 M HCl or HNO_3 for 30 min. Before use, the electrode should be soaked in the storage solution for at least an hour. The performance of a slow-responding electrode may be improved by sequentially soaking it for 10 min in 0.1 M HCl or HNO_3, 0.1 M NaOH, and finally, acid again. Before use, the electrode should be soaked in the storage solution for at least 1 h. In the case of protein and fat contamination, wiping the bulb gently with a tissue soaked in propanol or other alcohol should be enough to clean the glass bulb. Rinsing the electrode with mild detergent or methanol solution will also help. In the case of silicone contamination, a nitric acid soaking (5 M) for 2 h will dissolve the silicone.

15.5.2 Storage

An important part of maintenance is proper storage. For short-term storage (between measurements/up to 1 week), soak the electrode in the pH storage solution. This will provide the advantage that the experiment will always be ready to use. However, constant exposure of the electrode to the storage solution can

Table 15.2. Some Common Problems and Resolution with pH Measurement

Description	Possible Cause/Remedy
Drift	Liquid junction potential at the reference not constant/Change the reference electrode.
	Loose contact/Rectify the fault.
	Electrode not plugged in properly/Rectify the fault.
Sensitive to handling	Reference electrode is not filled/Top up with electrolyte solution, free of air bubbles.
	Reference electrodes filled with the wrong solution/Empty and refill the reference electrolyte.
	Diaphragm clogged/Clean diaphragm.
	Measurement of poorly conductive solutions/Measure with different amplifier or add supporting electrolyte.
Sluggish pH establishment	Adsorption at the glass membrane/Service glass membrane.
	Dirty diaphragm/Clean diaphragm.
Slope too low	Diaphragm contaminated/Clean diaphragm.
	Adsorption at glass membrane/Service glass membrane.
	Deswollen glass membrane after measurements in anhydrous solvents/Soak electrode in water between measurements.
	Old electrode/Regenerate glass membrane.
	Poor buffer solutions/Use fresh buffer solutions.
Slope cannot be adjusted	Diaphragm blocked/Clean diaphragm.
	Wrong order of buffer solutions/Use pH 7 buffer as the first buffer.
Shows the same value in pH 4 and 7 buffers.	Crack in the glass membrane/Replace the electrode.
	Connector damp or dirty/Dry and clean connector.

accelerate the aging process of the glass electrode. For long-term storage (more than 1 week), dry storage is preferred. The electrode can be used again after soaking in pH 7 buffer for a few hours. Unfortunately, the composition of the glass membrane will gradually deteriorate even during dry storage.

15.6 COMMON PROBLEMS AND SOLUTIONS

Common problems for pH measurements include potential or temperature drift, breakage of the electrode when it is handled, sluggish pH response, low value for the slope, and so on. Most pH meter manufacturers have provided quite intensive troubleshooting instructions in their instrument manual. Table 15.2 is a summary of some common solutions.

REFERENCES

1. I. Feldman, *Anal. Chem.*, **28**, 1861, 1956.
2. J. A. Illingworth, *Biochem. J.*, **195**, 259–262, 1981.
3. W. Davidson, and M. J. Gardner, *Anal. Chim. Acta*, **182**, 17–31, 1986.
4. R. W. Bates, *Determination of pH*, Wiley, New York, 1973.
5. S. L. Truman, Potentiometry: pH and ion-selective electrode, in W. E. Galen, *Analytical Instrument Handbook*, Marcel Dekker, New York, 1997.

16

QUALIFICATION OF ENVIRONMENTAL CHAMBERS

GILMAN WONG AND HERMAN LAM, PH.D.
GlaxoSmithKline Canada, Inc.

16.1 INTRODUCTION

An effective and compliant stability program will ensure that drug products developed for human use are safe, efficacious, and will meet a certain standard of quality (i.e., remain within its approved shelf life specification when exposed to various storage conditions). As the product may experience extreme temperatures and/or humidity as it moves through the supply chain from manufacturer to pharmacy to consumer, it is imperative to determine what effect these conditions will have on product quality. A number of systems are implemented to guarantee that the stability program is carried out with these principles in mind. The elements of an effective stability program include:

1. Policies and procedures
2. Test methods
3. Stability protocols
4. Stability samples
5. Environmental chambers
6. Sample tracking
7. Testing and evaluation of the stability data
8. Documentation

Analytical Method Validation and Instrument Performance Verification, Edited by Chung Chow Chan, Herman Lam, Y. C. Lee, and Xue-Ming Zhang
ISBN 0-471-25953-5 Copyright © 2004 John Wiley & Sons, Inc.

Each of these elements is a critical component of the stability program, but the proper operation of environmental chambers is one element that is most often assumed. It is often taken for granted that in the absence of external factors such as power loss or mechanical failure, a chamber, will maintain its predefined storage condition over the length of the stability study, but this is rarely the case. How do we ensure that the chamber will perform and maintain its set-point conditions as expected over the course of the stability study? A properly executed qualification program of the chamber prior to its routine use in a stability program will verify that the chamber can perform as expected.

Proper qualification of the environmental chambers will ensure that stability studies are performed at the proper storage conditions as required by national and regional regulatory authorities. It also provides assurance that the storage conditions within the chamber are uniform with regard to temperature, humidity, and any other predetermined parameters, such as light intensity level. The implementation of standardized storage conditions for the execution of stability programs as described by the International Conference on Harmonization (ICH) Q1 guidance documents for stability testing requires that proper chamber qualification be performed [1–3].

It is essential that an environmental chamber be qualified at its intended range of use prior to commissioning for routine operation in a stability program. A qualification protocol should be prepared describing the qualification procedures and must include predefined acceptance criteria for successful qualification. The qualification consists of three components: an installation qualification, operation qualification, and performance qualification.

16.2   INSTALLATION QUALIFICATION

Installation qualification (IQ) is a process used to ensure that the environmental chamber was delivered or built as ordered and is installed properly. The room where the chamber will be installed should have the proper utilities (e.g., power supply, feed water) and conditions (e.g., temperature) available to ensure that the chamber will operate according to the manufacturer's specifications during normal use. The review of the installation should include, but is not limited to, the following:

1. Startup checks
2. Utility checks
3. Component verification
4. Any additional IQ checks

The startup checks should be performed using a checklist that includes a record of the vital component information of the chamber, such as the model, type, serial number, and so on, of the following critical components:

1. Compressor
2. Chart recorder
3. System controllers
4. Steam generator
5. Dehumidifier

The startup checklist should also include a verification of the correct chamber set points, controller setup parameters, and compressor operating head pressure.

The utilities check is used to verify that the utility supplied fulfills the chamber requirements as specified by the manufacturer. During IQ, the "as found" parameters are verified against the "as specified" parameters on the checklist. If the "as found" results are significantly different from the "as specified" parameters, it will be necessary to determine the cause of this discrepancy and to implement corrective actions. The utilities check is a part of the site preparation for the installation of the chamber and should include verification of the power supply, such as the voltage, amperage, and wire size and the quality, pressure, and flow rate of the feed water supply. The quality of the feed water supply is a critical component of proper operation of the chamber. Experience has shown that if the quality of the feed water does not meet the manufacturer's specifications, this will lead to premature corrosion within the humidification and/or dehumidification system and subsequent problems with maintaining the humidity in the chamber within the specified limits.

During component verification, operation of the critical components of the chamber after delivery and installation is verified. These critical components include:

1. Compressor
2. Instrumentation, such as chart recorder, temperature and humidity sensors and controllers, and chamber alarms
3. Air handler
4. Steam generator
5. Dehumidifier

The manufacturer should verify the operation of these critical components during the installation according to their standard procedures for evaluating the proper operation of these components. Additional IQ checks will verify that the necessary documentation is available for proper operation of the chamber. These checks should confirm that standard operating procedures and equipment manuals are available. In addition, all calibrated instruments associated with the chamber should be documented in current calibration and placed on a calibration program.

16.3   OPERATION QUALIFICATION

Operation qualification (OQ) is a process used to ensure that the chamber components will operate according to predetermined specifications. OQ tests are designed to ensure that the chamber will function properly through a verification of all applicable operational features of the chamber. Only those aspects of the chamber that are deemed critical in maintaining the set-point conditions are tested.

The operation qualification should verify proper operation of the following components:

1. Controls and indicators, such as power, humidification/dehumidification, and refrigeration switches
2. Defrost timer
3. Temperature and humidity alarms

The chamber controls and indicators for the power, humidification/dehumidification, and defrost systems can be tested by turning them on and off and determining whether the indicator light is lit and the applicable system is activated. The defrost timer may be checked by activating the defrost switch and determining if the defrost cycle is initiated at the preset time. For the test of the chamber's alarms, a high- or low-temperature/humidity condition may be simulated through use of a calibrated standard attached to the chamber's alarm controllers. When the alarm condition is reached, the chambers audible alarm should be activated during the test. In addition, the chamber systems related to the alarm should be deactivated (e.g., a high-humidity alarm will deactivate the humidification system).

Temperature and/or humidity mapping of the chamber at the required set point is an integral part of OQ tests. Mapping includes the placement of temperature and humidity sensors at various locations within the chamber to demonstrate that the temperature and/or humidity will be consistent throughout different areas of the chamber under ideal conditions. Temperature and humidity readings should be obtained over a defined period, typically, a minimum of 24 h.

Sensors are calibrated against NIST-traceable standards prior to and after the completion of OQ testing. Thermocouples (TCs) are used to monitor temperature and are calibrated at three points: typically, $0°C$, the set-point temperature, and $50°C$ or above, depending on the set-point temperature. Thermocouples are temperature-sensing devices made by joining two dissimilar metals. This junction produces an electrical voltage in proportion to the difference in temperature between the hot junction (sensing junction) and the lead wire connection to the instrument (cold junction). Resistance-transmitting devices (RTDs), sensors that use the resistance temperature characteristic, are used to monitor humidity and are calibrated at two points, typically, 35% RH and 80% RH. Resistance-transmitting devices are more accurate than thermocouples because the signal from an RTD can be transmitted for long distances over wire and are not affected by voltage drops in the wire. In contrast, when signals are transmitted as a voltage, as in a thermocouple, the voltage drop caused by supply currents in the wires usually appears as an error in the signal measurement. As humidity fluctuates much more readily than temperature, it is more advantageous to use an RTD rather than a TC to perform the humidity portion of the mapping study. Temperature and/or humidity mapping is performed on an *empty* chamber because this would represent the ideal configuration of the chamber with no sample load to interfere with the airflow pattern within the chamber. Data are collected at 1-minute intervals.

Sensors are distributed equally in various areas of the stability chamber no less than 2 inches from any wall. A set of sensors should be placed near or at the temperature and/or humidity controller of the chamber, as the controller will maintain the set-point temperature and/or humidity within the chamber during normal use. For a typical walk-in chamber, a minimum of 24 thermocouples and six resistance-transmitting devices are recommended for use in the mapping study. For a benchtop or reach-in chamber, a reduced number of sensors may be used. It is important to note that regardless of the size of the chamber, the placement pattern of the sensors should be such that any potential hot or cold spots are mapped, particularly those areas near the door and corners of the chamber. Typical sensor placement patterns for a reach-in and walk-in chamber are shown in Figures 16.1 and 16.2, respectively. In these examples, the extremities of the chamber (i.e., top and bottom) have a larger number of sensors than the middle of the chamber, since these areas would have a greater probability of either hot or cold spots, due to the airflow pattern within the stability chamber.

The acceptance criteria for a successful operation qualification are based on the current ICH tolerance limits of $\pm2°C$ for temperature and $\pm5\%$ RH for humidity [6, 7]. As it is expected that the temperature and humidity uniformity of an empty chamber would be superior to that for a chamber containing product, the acceptance criteria for successful OQ is set tighter than the ICH tolerance limits to ensure that the chamber will meet the required specifications during normal use. Therefore, at each location within the stability chamber, it is recommended that the temperature and humidity should be within $\pm1°C$ and $\pm4\%$ RH of the predefined chamber set point.

16.4 PERFORMANCE QUALIFICATION

Performance qualification (PQ) is a process used to verify that the chamber performs according to specifications under normal operating conditions as defined by user requirements. PQ tests should include:

1. Temperature and humidity mapping under actual load conditions
2. Door-opening tests to determine chamber recovery time
3. Power-down test to determine how the temperature and humidity uniformity is affected during a power failure

The temperature and/or humidity mapping study for the PQ test is performed upon successful completion of the OQ test. It is similar to the OQ test; however, the test is performed under fully "loaded" conditions. The "load" in this case consists of empty containers used to simulate a fully loaded chamber. This would simulate a worst-case scenario where the chamber is fully loaded with product

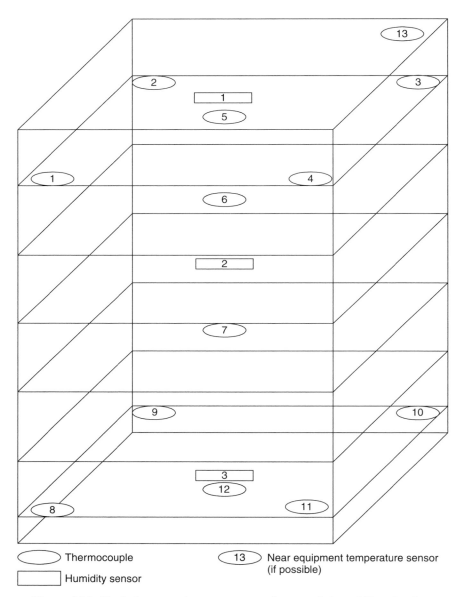

Figure 16.1. Typical sensor placement pattern for a reach-in stability chamber.

and there is a maximum disruption of airflow through the chamber. The mapping study is performed over a defined time interval, typically, a minimum of 48 h. The acceptance criteria are based on the ICH tolerance limits of $\pm 2^{\circ}$C for temperature and $\pm 5\%$ RH for humidity [6, 7]. At each location within the chamber, the temperature and/or humidity must be within the limits of the predefined set

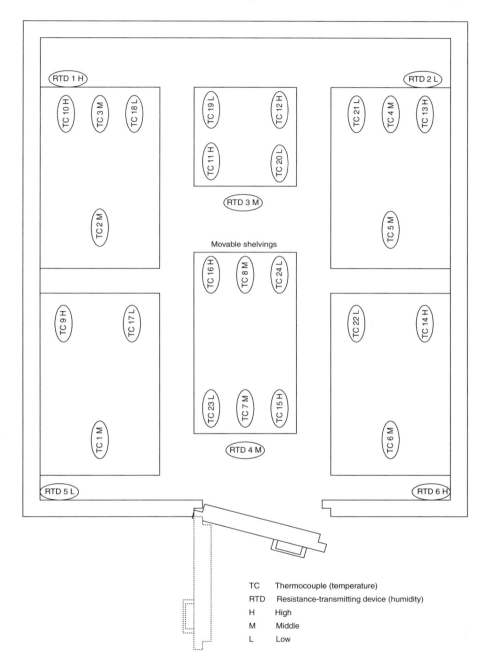

Figure 16.2. Typical sensor placement pattern for a walk-in stability chamber.

point. The placement of the temperature and humidity sensors is identical to that in the OQ test, and empty containers are placed in the chamber to disrupt the airflow.

The door test is performed at the completion of the temperature and/or humidity-mapping test under actual load conditions. The test is meant to simulate the opening of the door during sample retrieval from the chamber. The door is opened for a period of 1 min and then closed. The time for the chamber to recover to its set point ($\pm 2^\circ$C and $\pm 5\%$ RH) at all sensor locations is determined. The door test should be a minimum of 30 min in duration, with the chamber condition being recorded every minute during the test. Consideration should be made to perform this exercise a minimum of three times to determine an average recovery time. The information from the door test may be used to set the alarm delay for the chamber monitoring system and to provide information as to how long it takes for the chamber to recover to its set-point conditions after a routine door opening.

The power-down test is meant to simulate how temperature and humidity uniformity are affected during a power failure. During the test, the main power to the chamber is turned off and the conditions within the chamber are recorded every minute until the temperature and/or humidity exceeds the chamber set-point tolerance limits at every sensor location. The data collected from this study are used to determine how the chamber is affected by a power failure. It is also informative to continue to record the chamber conditions once the main power is restored to determine the length of time required for the chamber to recover to its set-point conditions.

Upon the completion of all qualification testing, a report should be written that summarizes the results of the testing. The minimum, maximum, and average values from each of the sensor locations should be provided and presented in graphical form. Any charts (Figure 16.3) and printouts generated during the qualification testing should be dated and signed and included in the qualification package. The instruments used in the testing should be included in the report, along with their calibration certificates where relevant.

In summary, a successfully executed qualification of a stability chamber will ensure that the chamber will perform within its intended range of operation and that stability studies are performed at the proper storage condition within specified tolerance limits and according to current regulatory expectations.

16.5 COMMON PROBLEMS AND SOLUTIONS

1. *What do you do if the acceptance criteria for successful qualification are not met: specifically, the limits for temperature and/or humidity?* Reasons for the failure to meet the acceptance criteria must be determined. There may be mechanical problems associated with the chamber that disrupt the airflow uniformity throughout the chamber. This will increase the likelihood of hot or cold spots developing in the chamber. This issue must be addressed before another qualification is attempted. If the problem cannot be resolved, a wider tolerance limit for normal

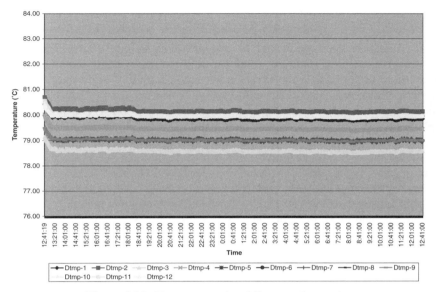

Figure 16.3. Temperature–humidity mapping study.

operation of the chamber at its predetermined set point must be established. It must be kept in mind that this would potentially limit use of the stability chamber for ICH stability studies.

2. *How often should chamber controllers be calibrated?* Chamber controllers should be calibrated a minimum of every six months. If controllers pass consecutive calibrations successfully, consideration should be given to extending the calibration interval.

3. *Should the chamber be revalidated routinely?* Consideration should be given to performing a revalidation of the chamber at a predefined interval (e.g., every two years). As the product load in the chamber changes during normal use, it may be necessary to revalidate the chamber. A performance qualification with the product providing the load would be sufficient for revalidation. If the chamber is monitored routinely at multiple locations for temperature and/or humidity, revalidation may not be necessary since a limited ongoing validation is being performed during normal use. In addition, if any critical chamber component is replaced (e.g., humidifier, controller), revalidation of the chamber by performing an OQ and/or PQ test may be required.

4. *Should the chamber be validated at a single set point or a set-point range?* If the set point of the chamber will not be changed, it may be sufficient to validate the chamber for use at a single set point. However, to provide additional flexibility, validating the chamber for use over a set-point range would be a good strategy, although this will involve more work at the start. By validating the chamber for use at the extremes of the set-point range, all set points within the

range are also validated for routine use. The set-point range selected should be based on the set points that may be needed for the chamber (e.g., 5 to 40°C).

5. *How is a chamber validated for photostability testing?* Validating a chamber for photostability testing is similar to that for temperature and/or humidity. In addition to confirming the temperature and/or humidity uniformity, the light output uniformity is also evaluated. Two types of light sources can be used for photostability testing according to existing ICH requirements [2]. An option 1 light source is designed to output both visible and UV light simultaneously; an option 2 system describes a system utilizing two sources, a cool fluorescent white source for the visible and a separate UV source. Light intensity is monitored using a lux meter (visible) and a spectroradiometer (UV). During validation, it is necessary to determine the light uniformity and establish acceptable limits. It is not unusual to have limits set at ±15% of the set point, and this is highly dependent on the type and make of light source that is used in the chamber.

6. *How are chamber deviations addressed during normal use?* Chamber deviations are investigated immediately and corrective actions taken to minimize the impact on the stability studies. As per current FDA and ICH requirements, study impact assessments are required for any chamber deviations that are greater than 24 h [1–4]. Depending on the length of the deviation, consideration should be given to extend the duration of the study to compensate for loss of exposure due to the deviation at the required storage condition as defined in the stability protocol. For example, if the product is to be stored for 30 days at 40°C/75% RH but a chamber deviation resulted in 6 days at 37°C/75% RH, the study should be extended an additional 6 days. Alternatively, a mean kinetic temperature (MKT) calculation may be performed to assess the overall temperature over the length of the study, taking the deviation into account. The MKT is defined as a single calculated temperature at which the total amount of degradation over a particular period is equal to the sum of the individual degradations that would occur at various temperatures. Thus, MKT may be considered as an isothermal storage temperature that simulates the nonisothermal effects of storage temperature variation [5]. The advantage of using a MKT calculation is that it will address temperature deviations both above and below the set-point. In addition, each impact assessment should be reviewed and approved by a quality representative.

7. *What about Preventive maintenance?* A preventive maintenance program should be established in consultation with those responsible for the upkeep of the chamber during normal use. This should be done after successful qualification and prior to releasing the chamber for use. Components that are prone to malfunction due to normal wear and tear include solenoid valves, humidifier floats, and recirculation-fan drive motors. These components should be checked or replaced on a regular basis to prevent or minimize chamber downtime.

8. *What needs to be done prior to executing the validation?* Prior to performing a validation, the validation protocol must be prepared and the acceptance criteria for successful validation should be established. The protocol must be reviewed and approved by the originating department as well as the engineering,

calibration, and quality assurance departments. Arrangements must be made to have the data recorder set up and calibrated prior to use in the validation study.

9. *How do you prepare for an audit of the environmental chambers?* In preparing for an audit, it is necessary that all validation documents be readily available for inspection. If these documents are stored off-site, arrangements should be made to retrieve these documents in time for the inspection. In addition, all chamber deviations should be identified and summarized in an annual report, if necessary. Also, the chamber inventory should be verified against existing records to ensure accuracy. An explanation should be available to explain any inventory discrepancies.

10. *How long does it take to validate a chamber for use?* It takes approximately two months to validate a chamber for routine use. Preparation of the validation protocol and subsequent review and approval takes the bulk of the time, or approximately one month. Actual testing will take about 2 weeks, and data analysis, report writing, and approval will take another 2 weeks.

11. *What about ongoing chamber monitoring and performance verification?* Environmental chambers should be monitored continuously by a central monitoring system. Consideration should also be given to monitoring larger environmental chambers such as walk-in chambers in several locations. This would provide an ongoing performance verification of the walk-in stability chamber, ensuring that the set-point conditions are maintained within the specified tolerance limits. Routine calibration of the sensors at predetermined intervals (e.g., every six months) ensures that readings within the chamber are accurate.

REFERENCES

1. ICH Harmonized Tripartite Guideline, ICH Q1A(R2): *Stability Testing of New Drug Substances and Products*, Feb. 2003.
2. ICH Harmonized Tripartite Guideline, ICH Q1B, *Photostability Testing of New Drug Substances and Products*, Nov. 1996.
3. ICH Harmonized Tripartite Guideline, ICH Q1C, *Stability Testing of New Dosage Forms*, Nov. 1996.
4. FDA Draft Guidance Document, *Stability Testing of Drug Substances and Drug Products*, June 1998.
5. *United States Pharmacopoeia*, USP 25, p. 2214, 2002.
6. ICH Harmonized Tripartite Web site, *www.ich.org*.
7. FDA Web site, *www.fda.gov/cder/guidance/index.htm*.

17

EQUIPMENT QUALIFICATION AND COMPUTER SYSTEM VALIDATION

LUDWIG HUBER, PH.D.
Agilent Technologies, Inc.

17.1   INTRODUCTION

Proper functioning and performance of equipment and computer systems play a major role in obtaining consistency, reliability, and accuracy of analytical data. Therefore, equipment qualification (EQ) and computer system validation (CSV) should be part of any good analytical practice. It is also requested by FDA regulations through the overall requirement that "equipment must be suitable for its intended use." While in the past, equipment qualification and computer validation focused on equipment hardware and stand-alone computer systems, recently the focus has been on network infrastructure, networked systems, and on the security, authenticity, and integrity of data acquired and evaluated by computer systems.

Because of their importance, validation issues have been addressed by several organizations and private authors:

1. The Pharmaceutical Analysis Science Group (UK) has developed a position paper on the qualification of analytical equipment [1].
2. The Laboratory of the Government Chemist (LGC) and Eurachem-UK has developed a guidance document with definitions and step-by-step instructions for equipment qualification [2].

Analytical Method Validation and Instrument Performance Verification, Edited by Chung Chow Chan, Herman Lam, Y. C. Lee, and Xue-Ming Zhang
ISBN 0-471-25953-5 Copyright © 2004 John Wiley & Sons, Inc.

3. The U.S. Food and Drug Administration (FDA) has developed principles of software validation [3] and computer systems [4] used in an FDA 21 CFR Part 11-regulated environment.

4. The Good Automated Manufacturing Practices (GAMP) Forum has developed guidelines for computer validation [5] and a draft guidance on quality assurance of information technology (IT) infrastructure [6].

5. Huber has published two validation reference books for the analytical laboratory [7,8]. The first one covers all validation aspects of an analytical laboratory, including equipment, analytical methods, reference compounds, and personnel qualification. The second covers the validation of computerized and networked systems.

6. The Parenteral Drug Association (PDA) has developed a technical paper on the validation of laboratory data acquisition system [9].

7. The Pharmaceutical Inspection Convention has developed a guideline on good practices for computers used in a regulated environment [10].

8. Huber has published primers with detailed information on the validation of macros and spreadsheet applications [11] and on the qualification of network infrastructure and the validation of networked systems [12].

All these guidelines and publications follow a couple of principles:

1. Qualification of equipment and validation of computer systems are not one-time events. They start with the definition of the product or project and setting user requirement specifications and cover the vendor selection process, installation, initial operation, ongoing use, and change control.

2. All publications refer to some kind of life-cycle model, with a formal change control procedure being an important part of the entire process.

3. There are no detailed instructions on what should be tested. All guidelines refer to some kind of risk assessment for the extent of validation.

17.2   SCOPE OF CHAPTER

In this chapter we guide quality assurance (QA) managers, lab managers, IT personnel, and users of equipment hardware and software through the entire qualification and validation process, from writing specifications and vendor qualification to installation and initial and ongoing operation. The following points are covered:

1. Qualification of equipment hardware (e.g., a spectrophotometer or liquid chromatograph).

2. Qualification of computer hardware with peripherals and accessories such as printers and disk drives.

3. Validation of software loaded on a computer that is used to control the equipment, to acquire analytical raw data, to process the data, and to print and store the data. Software typically includes operating systems, standard applications software, and software written by a specific user, such as macros or spreadsheet calculations.

4. Qualification of network infrastructure and validation of networked systems.

5. Validation of functions that are required to meet electronic records and signature compliance, such as 21 CFR Part 11.

6. Documentation as required by regulations.

In Section 17.4 we discuss common problems and offer solutions.

Throughout the chapter we take into account most national and international regulations and quality standards. The concept, examples, templates, and operating procedures described are based on the author's multinational experience and incorporate all aspects of validation and qualification used at Agilent Technologies and taken from personal discussions with regulatory agencies, managers and chemists in laboratories, corporate QA and IT managers, and vendors of equipment and software. Readers will learn how to speed up their validation and qualification process, thereby avoiding troublesome reworking, and will gain confidence for audits and inspections.

The validation and qualification procedures presented in this chapter help to assure compliance and quality at minimal extra cost and minimal administrative complexity. The purpose is to answer the key questions regarding validation: How much validation is needed, and how much is sufficient? The recommendations are complementary rather than controversial to any standards or official guidelines. They are based mainly on common sense and can be used in cases where information from official guidelines and standards is insufficient for day-to-day work. We list the major steps of each qualification, but for an in-depth understanding and for easy implementation, readers are directed to the literature [2,7,8,13].

17.3   EQUIPMENT QUALIFICATION AND VALIDATION OVERVIEW

Equipment qualification covers the entire life of a product. It starts when somebody has an idea about a product and ends when the equipment is retired. For computer systems validation, this ends when all records on the computer system have been migrated and validated for accuracy and completeness on a new system. Because of the length of time and complexity, the process has been broken down into shorter phases: design qualification (DQ), installation qualification (IQ), operational qualification (OQ), and performance qualification (PQ) [2]. The process is illustrated in Figure 17.1.

Validation activities should be described in a validation master plan which provides a framework for a thorough and consistent validation. Regulatory agencies typically do not specifically demand a validation master plan. However, inspectors want to know the company's approach toward validation. The validation

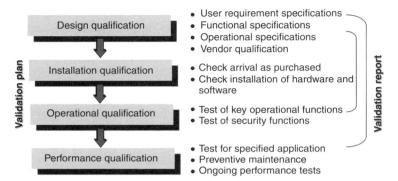

Figure 17.1. Equipment qualification phases. (With permission from Ref. [7].)

master plan is an ideal tool to communicate this approach both internally and to inspectors. It also ensures consistent implementation of validation practices and makes validation activities much more efficient. In case there are any questions as to why things have been done or not done, the validation master plan should provide the answers.

Within an organization, a validation master plan can be developed for:

- Multiple sites
- Single site
- Single location
- Single system category
- Department categories (e.g., for quality assurance departments)

Equipment and computer validation master plans should include the following:

1. Introduction with a scope of the plan (e.g., sites, systems, and processes)
2. Responsibilities
3. Related documents (e.g., validation policy)
4. Products/processes to be validated and/or qualified
5. Validation approach
6. Steps for equipment qualification and computer system validation with examples on type and extent of testing
7. Handling of the existing systems
8. Validation of macros and spreadsheet calculations
9. Change control procedures and templates
10. Instrument obsolescence and removal
11. Training plans (system operation, GMP)
12. Templates and references to SOPs
13. Glossary

For larger projects, an individual validation project plan should be developed. This plan is derived from the validation master plan. It formalizes qualification and validation and outlines what has to be done to get a specific system into compliance. For inspectors it is the first indication on the control that a laboratory has over a specific computerized system, and it also gives a first impression of the validation quality.

A validation project plan should include the following sections:

1. Scope of the system: what it includes and what it does not include
2. System description
3. Responsibilities
4. Test strategy and approach for DQ, IQ, OQ, PQ
5. Acceptance criteria
6. Ongoing validation
7. System retirement
8. References
9. Time line and deliverables for each phase

17.3.1 Design Qualification

"Design qualification (DQ) defines the functional and operational specifications of the instrument and details the conscious decisions in the selection of the supplier" [2]. DQ should ensure that instruments have all the necessary functions and performance criteria that will enable them to be implemented successfully for the intended application and to meet business requirements. Errors in DQ can have a tremendous technical and business impact, and therefore a sufficient amount of time and resources should be invested in the DQ phase. For example, setting the wrong operational specifications can substantially increase the workload for OQ testing, and selecting a vendor with insufficient support capability can decrease instrument up-time with a negative business impact. The following steps are recommended for inclusion in a design qualification.

- Description of the analytical problem and selection of the technique
- Description of the intended use of the equipment
- Description of the intended operation environment (includes the computer environment)
- Preliminary selection of the functional and performance specifications (technical, environmental, safety)
- Preliminary selection of the supplier
- Instrument tests (if the technique is new)
- Final selection of the equipment, functional and performance specifications
- Final selection and qualification of the supplier

- Development and documentation of final functional and operational specifications

To set the functional and performance specifications, the vendor's specification sheets can be used as guidelines. However, it is not recommended that the vendor's specifications simply be copied because compliance to the functional and performance specifications must be verified later, during operational qualification and performance qualification. Specifying too many functions and/or setting the values too stringently will significantly increase the workload for OQ.

17.3.2   Installation Qualification

Installation qualification (IQ) establishes that the instrument is received as designed and specified, that it is properly installed in the selected environment, and that this environment is suitable for the operation and use of the instrument. The following steps are recommended before and during installation.

Before Installation

1. Obtain the manufacturer's recommendations for installation site requirements.
2. Check the site for fulfillment of the manufacturer's recommendations (utilities such as electricity, water, and gases, and environmental conditions such as humidity, temperature, vibration level, and dust).
3. Allow sufficient shelf space for the equipment, SOPs, operating manuals, and software.

During Installation

1. Compare equipment, as received, with the purchase order (including software, accessories, and spare parts).
2. Check the documentation for completeness (operating manuals, maintenance instructions, standard operating procedures for testing, safety, and validation certificates).
3. Check the equipment for damage.
4. Install the hardware (computer, equipment, fittings, and tubings for fluid connections, columns in HPLC and GC, power cables, data flow, and instrument control cables).
5. Switch on the instruments and ensure that all modules power up and perform an electronic self-test.
6. Install the software on the computer following the manufacturer's recommendation.
7. Verify correct software installation (e.g., verify that all files are loaded). Utilities to do this should be included in the software itself.
8. Make a backup copy of the software.

9. Configure the peripherals (e.g., printers and equipment modules)
10. Identify and make a list with a description of all hardware; include drawings where appropriate.
11. Make a list with a description of all software installed on the computer.
12. List equipment manuals and SOPs.
13. Prepare an installation report.

17.3.3 Operational Qualification

"Operational qualification (OQ) is the process of demonstrating that an instrument will function according to its operational specification in the selected environment" [2]. Before OQ testing is done, one should always consider what the instrument will be used for. There must be a clear link between the OQ and DQ phase. Testing may be quite extensive if the instrument is to be used for all types of applications and some of these applications put high demands on the performance of the system. For example, if a chromatograph is intended for use with certain applications that work at low limits of quantitation (LOQ) and also with other applications that require quantitation of large amounts of analytes, the instrument's capability to quantitate trace and high levels of analytes should be verified. In this case we recommend using generic standards (e.g., caffeine) that will test the instrument for its general-purpose application but not for a specific application. On the other hand, if the instrument is to be used for one application only, the tests and acceptance criteria should be limited to that application. In this case, the test compound can be the same as the compounds analyzed during routine use.

If a system comprises several modules, it is recommended that system tests be performed for parameters that are affected by multiple modules (holistic testing) rather than performing tests module by module (modular testing). Individual module tests should be performed if the parameter is affected by that module only (e.g., the wavelength accuracy of an HPLC variable-wavelength detector or the temperature accuracy of a column compartment).

OQ tests are very commonly used in the ongoing performance verification or calibration of the system. Test templates should be used for OQ. An example is shown in Figure 17.2. Most important is that acceptance criteria should be defined before the test, and both acceptance criteria and test results should be documented in the test sheet.

The frequency of OQ/performance verification depends not only on the type of instrument and the stability of the performance parameters, but also on the acceptance criteria specified. In general, the time intervals should be selected such that the probability is high that all parameters are still within the operational specifications. Otherwise, analytical results obtained with that particular instrument are questionable. The OQ/performance verification history of the type of instrument can be used to set reasonable test intervals. Here the importance of proper selection of the procedures and acceptance limits becomes very apparent.

Test number:
Specification:
Purpose of Test:
Test Environment (PC hardware, peripherals, interfaces, operating system, Excel version, service pack):
Test execution: Step1: Step 2: Step 3:
Expected result: Acceptance criterion: Actual result: Comment:
Criticality of test: Low   0   Medium 0   High 0
Test Person:
Printed name: _____   Signature: _____   Date: _____

Figure 17.2. Example of a test script. (With permission from Ref. [11].)

For example, if the baseline noise of a UV–Vis detector is set to the lowest possible limit of the specification, the lamp will have to be changed more frequently than if it is set five times higher.

17.3.4   Performance Qualification

"Performance qualification (PQ) is the process of demonstrating that an instrument consistently performs according to a specification appropriate for its routine use" [2]. Important here is the word *consistently*. Important to consistent equipment performance are regular preventive maintenance, system changes in a controlled manner, and regular testing. The PQ test frequency can be much higher than for OQ. Another difference is that PQ should always be performed under conditions that are similar to routine sample analysis. For a chromatograph, this means using the same column, the same analysis conditions, and the same or similar test compounds.

PQ should be performed on a daily basis or whenever the instrument is used. The test frequency not only depends on the stability of the equipment but on everything in the system that may contribute to the results. For a liquid chromatograph, this may be the chromatographic column or a detector's lamp. The test criteria and frequency should be determined during the development and validation of the analytical method.

In practice, PQ can mean system suitability testing, where critical key system performance characteristics are measured and compared with documented, preset limits. For example, a well-characterized standard can be injected five or six times and the standard deviation of amounts are then compared with a predefined limit. If the limit of detection and/or quantitation are critical, the lamp's intensity profile or the baseline noise should be tested. Following are the steps recommended for (system suitability) PQ tests.

1. Define the performance criteria and test procedures.
2. Select critical parameters and acceptance limits. For a liquid chromatography system, it can be the following parameters:
   - Precision of the amounts
   - Precision of the retention times
   - Resolution between two peaks
   - Peak width at half height or peak tailing
   - Limit of detection and limit of quantitation
   - Wavelength accuracy of a UV–Vis wavelength detector
3. Define the test intervals; for example:
   - Every day
   - Every time the system is used
   - Before, between, and after a series of runs
4. Define corrective actions on what to do if the system is out of specification

The analysis of quality control (QC) samples with the construction of quality control charts has been suggested as another way of performing PQ. Control samples with known amounts are interdispersed among actual samples at intervals determined by the total number of samples, the stability of the system, and the precision specified. The advantage of this procedure is that the system performance is measured more or less continuously under conditions that are very close to the actual application.

A frequently discussed question is whether either system suitability testing or the analysis of QC samples are sufficient to prove ongoing system performance, or whether additional checks should be performed. The answer to this question depends very much on the conditions under which the control samples are analyzed. For example, if the system is used for trace analysis and the amounts of the control sample do not include trace-level amounts, the capability of the system to measure low amounts should be verified. In HPLC, this could be a routine check of the wavelength accuracy, the baseline noise, and the intensity of the UV lamp. System suitability checks and control sample analysis are sufficient as PQ checks if all critical system parameters are checked as part of tests and evaluations.

17.3.5 Validation of Macros and Spreadsheet Calculations

Macro programs are popular in analytical laboratories for evaluating analytical data and deriving characteristics of the products analyzed. Spreadsheets are used widely in laboratories in nontrivial activities as diverse as job processing, data capture, data manipulation, and report generation. Databases are used to correlate data from a single sample analyzed on different instruments and to obtain long-term statistical information for a single sample type. The processes may be automated using macros: for example, enabling the analytical data to be transferred, evaluated, and reported automatically. In all these programs, analytical data are converted using mathematical formulas.

Today the understanding is that the programs themselves do not have to be validated by the user (e.g., MS Excel). What should be validated are the calculations and program steps that a user has written. There should be documentation on what the application program written by the user as an add-on to the core software is supposed to do, who defined and entered the formulas, and what the formulas are. Validation activities should follow a life-cycle model starting with defining user requirements, followed by design specification, code development, and structural and functional testing. For small projects some of the typical phases, such as functional testing and testing in a user's environment, can be combined (see Figure 17.3). It is of utmost importance that changes follow standard procedures for initiation, authorization, implementing, testing, and documenting. All activities should be planned in the validation project plan and documented in the validation report.

The programmer and later, anticipated users should test and verify the functioning of the program. A frequently asked question is: How much testing should be conducted? Testing should demonstrate that the system is providing accurate, precise, and reliable results. A spreadsheet, for example, may be tested using typical examples throughout the anticipated operating range. Results obtained by the program should be compared with results obtained by a calculator. If the spreadsheet is to be used to further evaluate small numbers, the test program should also include small numbers. If an accurate calculation requires three digits after the decimal, the test should use numbers with at least three digits after the decimal. The test program should also include stress testing with values outside

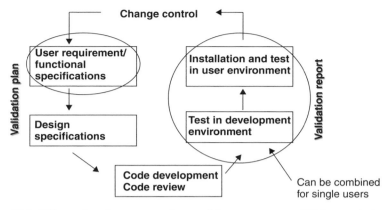

Figure 17.3. Life cycle of macros and spreadsheet programs. (With permission from Ref. [11].)

the normal operation range. Test cases should be linked and traceable to user requirements and functional specifications in a traceability matrix.

Test protocols should be used with information on the following subjects:

- Functions that should be tested
- Steps to be performed
- Equipment used
- Data inputs and expected output
- Acceptance limit
- Name of test person
- Date of testing

The tests should be repeated thereafter at regular or nonregular intervals. Therefore, any test case should be designed for reuse using sets of data inputs and known outputs. If the program is used infrequently, it is good practice to run the tests each time the program is used. Spreadsheet validation is discussed in detail in Chapter 18 and in Ref. [11].

17.3.6   Change Control

Any changes to specifications and programming codes should follow written procedures and be documented. Changes may be initiated because errors have been found in the program or because additional or different functions may be desirable. Requests for changes should be submitted by users and authorized by the user's supervisor or department manager and by the quality assurance department. After any changes the program should be tested. Full testing should be done for the part of the program that has been changed and regression testing

Form ID:	
System ID:	
System Location:	
Change Initiator:	
Description of change (should include reason for change and business benefit):	
Expected impact on validation:	
Authorization to change:	Name: Signature: Date:
Change implemented on:	Date:
Comments (implementation, testing):	e.g., document any observation and new version or revision number, and types of tests that have been performed.
Completed by:	Name: Signature: Date:
Approved by:	Name: Signature: Date:

Figure 17.4. Template for change control.

should be done for the entire program. An example template for change control is given in Figure 17.4.

17.3.7 Qualification of Network Infrastructure and Validation of Networked Systems

Networked systems with integrated databases are used extensively in the pharmaceutical industry. These are computerized systems, and as such, must be qualified and validated to demonstrate suitability of their intended use. While validation of stand-alone computer systems is well described [8] and understood, there is still uncertainty on how to qualify networks and networked systems. On the other hand, inspectors are looking more and more into such systems, and validation of such systems is important for business reasons. For example, missing data in a laboratory information system or lost data from a research project can be disaster for a company and for people individuals in the company. The same holds for production delays caused by network failures.

Quality assurance of networks had been addressed by Crosson et al [14]. They recommended that a network should be qualified as if it was a piece of equipment and then managed in a documented state of control. Network quality was also covered by a Special Interest Group (SIG) of the Good Automated Manufacturing Practice (GAMP) Forum [6]. The document emphasizes quality assurance principles as being critical to management of the IT infrastructure. The group recommends bringing IT infrastructure into initial compliance with established standards through a planned qualification process. Once in compliance, the infrastructure should be kept in this state by a program of documented standard processes and quality assurance activities. The effectiveness of this program should be monitored periodically by an audit.

Figure 17.5 shows a typical client–server networked system connecting client computers in a laboratory and office computers to a server located in a computer room. The computer room also hosts mail servers. The laboratory computers, with data system applications software, control equipment with built-in local area network (LAN) cards that acquire data using TCP/IP protocols. Application software on these client computers is also used for data evaluation. Computers are connected to a server computer through a hub. The server has a relational database from Oracle (with customized applications for data management, control charting, and other statistical evaluation) for generating electronic signatures and for review, backup, archiving, and retrieval of data.

Validating networked systems requires qualifying individual network components: for example, applications running on each computer. It also means qualifying authorized access to the networked system and qualifying data transfer between two computers, which qualifies the interfaces of components at both sites. The entire system is validated by running typical day-by-day applications under normal and worst-case conditions and verifying correct functions and performance with criteria specified previously. For both qualification of the components and validation of complete systems, it is important to define a validation box. This box should define which parts of the complete network need to be qualified and which are not affected. In the example shown in Figure 17.5, the validation box for the laboratory data system includes lab computers, file server, applications server, and the database. Limiting the network qualification tasks to the components used by network applications can save time in the testing.

The two steps for complete network qualification and system validation are described briefly below [12].

Steps to Build a Qualified Network Infrastructure

- Specify network requirements. Specifications should include network devices, software, computer hardware, and computer peripheral cables. Specifications are based on anticipated current and future use of the network.

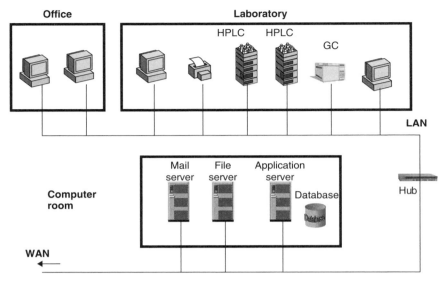

Figure 17.5. Example for a client–server networked system. (With permission from Ref. [12].)

- Develop a network infrastructure plan.
- Design network infrastructure and drawings.
- Select equipment and vendors for computers, NOS, network devices, and so on.
- Order equipment: computer hardware, software (OS, NOS), network devices, peripherals.
- Install all hardware devices according to design drawings and vendor documentation.
- Perform self-diagnostics, document hardware installation, and settings (this completes the IQ part).
- Document this as the network baseline.
- Make a backup of installed software and network configurations. Whatever happens, it should be possible to return to this point.
- Test communication between networked computers and peripherals, and access control, including remote access control.
- Develop and implement rigorous configuration management and change control procedure for all your network hardware and software. This should also include updates of system drawings if there are any changes.
- Before applying any system changes to a production environment, they should be verified in a test environment to ensure that one does not affect the intended functionality of the system.
- Monitor ongoing network traffic using a network health monitoring software.

Steps to Define, Implement, and Validate Applications Running on a Qualified Network

- Develop a validation plan and schedule using your validation master plan as a guideline.
- Specify applications software that runs on the qualified network (e.g., networked data system)
- Select and qualify the vendor.
- Install application software and perform IQ (set up, define, and document configuration settings, verify "proper" software installation through installation verification routines).
- Verify correct functioning of the application software. Apply common computer validation practices.
- Testing should include network transactions under normal and high loads. For this test, refer to verifications in the TCP/IP transfer protocol. The advantage here is that this is built into the system and will be performed on an ongoing basis. However, this is not 100% accurate, as there are rare situations where the test does not work. To be on the safe side, use routines (e.g., MD5 hash calculation routines based on 128-bit strings). Ask the vendor of your application software; sometimes, such tests are built in.
- Monitor ongoing performance of your application. The type of performance test will depend on the application.
- Monitor network connections and traffic using network health monitoring software.

As part of the installation system, drawings and diagrams should be generated. They are an absolute must not only for setting up a network and networked systems, but even more important, for maintaining them. They should be part of the IQ documents and should include the following diagrams:

- Physical diagrams such as component locations and cable network.
- Logical diagrams (e.g., TCP/IP schemes and how components interrelate with each other).
- If a dynamic IP addressing is used in the network scheme, a procedure should also be in place to indicate how this dynamic IP addressing is being utilized (including the procedure for submasking of the IP addresses). This will enable appropriate tracing of data or traffic flow in case such a tracing is needed to prove the data integrity and security.

Since networks change frequently, maintaining these diagrams with documented version control is important. A good recommendation is to have procedures in place for regular review of these diagrams (e.g., quarterly).

17.3.8 Validation Compliance with Electronic Records and Signatures

In 1997, the U.S. Food and Drug Administration (FDA) issued a regulation that provides criteria for acceptance by the FDA of electronic records, electronic signatures, and handwritten signatures [15]. This was done in response to requests from the industry. With this regulation, entitled Rule 21 CFR Part 11, electronic records can be equivalent to paper records and handwritten signatures. The rule applies to all industry segments regulated by the FDA that includes Good Laboratory Practice (GLP), Good Clinical Practice (GCP), and current Good Manufacturing Practice (cGMP). The use of electronic records is expected to be more cost-effective for the industry and the FDA. The approval process is expected to be shorter, and access to documentation will be faster and more productive.

The primary requirements of the new regulation for the analytical laboratories are as follows:

- Use validated existing and new equipment and computer systems.
- Secure retention of electronic records to reconstruct the analysis instantly.
- User-independent computer-generated time-stamped audit trails.
- System and data security, data integrity, and confidentiality through limited authorized system access.
- Use of secure electronic signatures for closed and open systems.
- Use of digital signatures for open systems.

It is beyond the scope of this chapter to discuss and explain how the requirements can be implemented in analytical laboratories. This has been described in a six-article series published in *Biopharm* [16–21]. We elaborate here on the validation aspect of the rule. Part 11 requires that computer systems used to acquire, evaluate, transmit, and store electronic records should be validated. This is not new, as processes and steps to validate such systems were described earlier in the chapter. FDA's expectations for validation have been described in the Part 11 draft guidance on validation [4]. This guidance makes it very clear that functions as required by Part 11 should be validated in addition to functions that are required to perform an application such as chromatographic instrument control, data acquisition, and evaluation. Specific functions as required by Part 11 are as follows:

- Limited authorized access to data and systems
- Computer-generated audit trail
- Binding signatures to records
- Accurate copies of electronic records
- Ready retrieval for processing of data

Required steps to achieve validation compliance are not different from other validation steps:

- Specify requirements and include them in your user requirement specifications document. Information on the requirements can be obtained from the regulation itself and from Part 11 industry guidance documents.
- Develop test procedures to verify that the functions meet the requirements.

Recommended test procedures include:

1. *Validation.* Part 11 validation requirements are the same as these described earlier in this chapter. One should validate not only functions that are required by the application but also specific functions, as required by Part 11.
2. *Limited and authorized system access.* This can be achieved by entering correct and incorrect password combination and verify if the system behaves as intended.
3. *Limited access to selected tasks and permission.* This can be achieved by trying to access tasks as permitted by the administrator and verify that the system behaves as specified.
4. *Computer-generated audit trail.* Perform actions that should go into the e-audit trail according to specifications. Record the actions manually and compare the recordings with a computer-generated audit trail.
5. *Accurate and complete copies.* Calculate an analytical result from the raw data using a defined set of evaluation parameters (e.g., integrator events and calibration tables). Save raw data, final results, and evaluation parameters on a storage device. Switch off the computer. Switch it on again and perform the same tasks as before using data stored on the storage device. Results should be the same as the original evaluation.
6. *Binding signatures with records.* Sign a data file electronically. Check the system design and verify that there is a clear link between the electronic signature and the data file. For example, the link should include the printed name or a clear reference to the person who signed, the date and time, and the meaning of the signature.

17.4 COMMON PROBLEMS AND SOLUTIONS

In this section we describe problems that users of equipment and computer frequently experience during validation and qualification. Problems have been either reported to the author during seminar or conference discussions or have been documented by the FDA in inspection reports and warning letters [22]. In this section we also give recommendations on how to avoid the problems or to resolve them if they occur.

1. *Specifications are missing or are not followed.* The most important part of validation is to write specifications. They include user requirement specifications

and functional specifications that can also be combined into a system requirement specifications document. Validation is impossible without clear specifications. One of the first requests during an audit is the request for specifications. The recommendation is to spend more time in defining the system requirements and its functional specifications. Specifications should be linked to test cases.

2. *Vendor specifications are used as acceptance criteria for operational qualification of equipment.* Vendors typically define specifications for analytical equipment instruments such that they can be met easily at the time of installation. However, these specifications are so stringent that problems arise after a period of usage of the instruments, and requalification tests fail.

Let us take a closer look at the definition of OQ and select the appropriate acceptance criteria. OQ is defined by the Eurachem/LGC (2) as "The process of demonstrating that an instrument will function according to its operational specification in the selected environment." A similar definition for OQ came from the U.S. Pharmaceutical Manufacturers Association (PMA) [23]: "Documented verification that the equipment related system or subsystem performs as intended throughout representative or anticipated operating ranges." Even though the definition is short and leaves a lot of room for interpretation, one thing becomes obvious: OQ should prove that the instrument is suitable for its intended use. OQ is not required to prove that the instrument meets the manufacturer's performance specifications. This is a frequent misunderstanding, and many operators prefer to use the manufacturer's specifications because these are usually readily available. However, a mistake such as this can have an enormous impact on the equipment's maintenance costs. One example is the baseline noise of an HPLC UV–Vis detector: a performance criterion that is important for limit of detection and limit of quantitation.

The baseline noise as offered by many UV–Vis detectors is in the range 1 to 2×10^{-5} AU and much lower than the limit of detection and quantitation required for most applications. This value is achieved under optimum conditions, such as with a reasonably new lamp, an ultraclean flow cell, stable ambient temperature, HPLC-grade solvents, and no microleaks in the entire HPLC system. These conditions are always valid at the manufacturer's final test and probably at the time of installation in the user's laboratory. However, after some time, optical and mechanical parts deteriorate (e.g., the lamp loses intensity and the flow cell may become contaminated). If we repeat the test after 3, 6, or 12 months, the noise of 1×10^{-5} AU may no longer be obtained. The recommendation is to select acceptance criteria according to the intended use of the system.

3. *Too much or not enough testing is done.* A frequent question is: How much testing is enough? Too much testing adds incremental workload and unnecessary cost. Insufficient testing can result in poor analytical data and inspectional observations and warning letters. There are no specific guidelines from any regulatory agency. The frequent answer to this question from experts is: It depends. FDA's Part 11 validation guidance [4] has a paragraph on commercial off-the-shelf systems that says: "All functions that the end user will use should be tested." However, in another paragraph it states:

"When you determine the appropriate extent of system validation, the factors you should consider include (but are not limited to) the following:

- The risk that the system poses to product safety, efficacy, and quality
- The risk that the system poses to data integrity, authenticity, and confidentiality
- The system's complexity; a more complex system might warrant a more comprehensive validation effort"

This approach has been confirmed by FDA's guidance on scope and applications. It requires justifying and documenting the rationale behind what and how much to validate: "We recommend that you base your approach on a justified and documented risk assessment and a determination of the potential of the system to affect product quality and safety, and record integrity." [24]

The recommendation is to look at all requirements in the user requirement specifications or system requirement specifications and define the ones that should be tested. Criteria are the criticality of the functions for the data accuracy and integrity of data generated or evaluated by the equipment hardware and software. This can be described as a yes/no statement in a row named 'testing required' if the rationale is self-explainable through common sense and good understanding of the product and its functions. For a chromatographic systems, critical functions are instrument control, data acquisition, peak integration, quantitation, and file storage and retrieval. A company's validation master plan should give guidelines and examples for the type of testing. This ensures consistent practice through an organization.

4. *Change control is absent or inadequate.* Most errors are introduced into a system after changes are made without authorization, documentation, and revalidation. For example, FDA's principles of software validation [3] state in Paragraph 2.4: "Of those software related recalls, 79% were caused by software defects that were introduced when changes were made to the software after its initial production and distribution." 21 CFR Part 11 validation guidance also states the following:

- Contractor or vendor upgrades or maintenance activities, especially when performed remotely (i.e., over a network), should be monitored carefully because they can introduce changes that might otherwise go unnoticed and have an adverse effect on a validated system.
- Examples of such activities include installation of circuit boards that might hold new versions of "firmware," software, addition of new network elements, software "upgrades," "fixes," and "service packs."
- It is important that system users be aware of such changes to their system. You should arrange for service providers to provide advice regarding the nature of revisions to assess the changes and perform appropriate revalidation. The recommendation is to control and document any change to a system, no matter whether it was done by the user or the vendor and to assess the need for and extent of revalidation. Vendors should be informed

about the user firm's change control procedures. Vendors should provide information on the type of change and on the impact the change has on data integrity and accuracy. This is equally important for changes to equipment hardware, firmware, or software.

5. *Documentation must be complete.* On completion of equipment qualification and system validation, documentation should be available that consists of the following:

- Validation master plan
- Validation project plan
- Design qualification document consisting of user requirement specifications (URS) and functional specifications (FS) [URS and FS can be combined into one system requirement specifications document (SRS)]
- Vendor assessment
- Installation qualification document (For computer systems, this should include an installation verification test report)
- For networked systems: network diagrams
- Entries of instrument identity in the laboratory instrument database
- Test plan with test traceability matrix
- Procedures for testing with acceptance criteria
- Qualification test reports with signatures and dates
- List of authorized users (the list should be signed and reviewed regularly)
- Procedures for preventive maintenance
- Change control procedures and change logs
- PQ test procedures and representative results
- Validation summary report

17.5 CONCLUSIONS

The most important factors for the entire process of equipment qualification and computer system validation in analytical laboratories are proper planning, execution of qualification according to the plan, and documentation of the results. The process should start with the definition of the analytical technique and the development of user requirement and functional specifications. For computer systems, a formal vendor assessment should be made. This can be done through checklists and vendor documentation with internal and/or external references. For very complex systems, it should go through a vendor audit.

The installation should be documented. The accuracy of software installation should be verified, and for networked systems, drawings with diagrams should be generated. The instrument should be tested for compliance to user requirements and functional specifications, as defined during the design qualification. Critical parameters should be tested before and during routine analysis. System suitability testing and the analysis of quality control samples, together with additional tests of critical parameters, are suitable tools for this ongoing performance qualification. All changes to the system should follow a documented change control procedure. These changes should be authorized before implementation and a formal assessment should be made on the need and extent of revalidation.

REFERENCES

1. M. Freeman, M. Leng, D. Morrison, and R. P. Munden from the UK Pharmaceutical Analytical Sciences Group (PASG), Position paper on the qualification of analytical equipment, *Pharm. Technol. Eur.*, pp. 40–46, Nov. 1995.

2. P. Bedson and M. Sargent, The development and application of guidance on equipment qualification of analytical instruments, *Accredit. Qual. Assur.*, **1**(6), 265–274, 1996.

3. U.S. Food and Drug Administration (FDA), *General Principles of Software Validation: Final Guidance for Industry and FDA Staff*, Rockville, MD, Jan. 2002.

4. U.S. Food and Drug Administration (FDA), *Guidance for Industry, 21 CFR Part 11; Electronic Records; Electronic Signatures Validation*, Rockville, MD, Aug. 2001.

5. *GAMP (Good Automated Manufacturing Practice) Guide for Validation of Automated Systems in Pharmaceutical Manufacturing*, Version 3, Mar. 1998, Version 4, Dec. 2001.

6. GAMP Special Interest Group (SIG), *IT Infrastructure: Quality Assurance* (draft), Nov. 2000.

7. L. Huber, *Validation and Qualification in Analytical Laboratories*, Interpharm Press, Englewood, CO, Oct. 1998.

8. L. Huber, *Validation of Computerized Analytical and Networked Systems*, Interpharm Press, Englewood, CO, Apr. 2002.

9. Parenteral Drug Association (PDA), *Validation and Qualification of Computerized Laboratory Data Acquisition Systems (LDAS)*, Technical Paper 31, 2000.

10. Pharmaceutical Inspection Convention, *Good Practices for Computerised Systems in Regulated "GxP" Environments* (draft), Jan. 2002.

11. L. Huber, *Using Macros&Spreadsheets in a Regulated Environment*, part of the Macro& Spreadsheet quality package, Labcompliance, *www.labcompliance.com/books/macros*, May 2002.

12. L. Huber, *Using Networks in a Regulated Environment*, part of the Network quality package, Labcompliance, *www.labcompliance.com/books/network-quality.htm*, Apr. 2001.

13. Tutorial on equipment qualification, *www.labcompliance.com/tutorial/equipment*.

14. E. Crosson, M. W. Campbell, and T. Noonan, Networking management in an FDA regulated environment, *PDA J.*, **53**(6), 280–287, Nov.–Dec. 1999.

15. *Code of Federal Regulations*, Title 21, *Food and Drugs*, Part 11, Electronic Records; Electronic Signatures; Final Rule, *Fed. Reg.*, **62**(54), 13429–13466.

16. L. Huber, Implementing 21CFR Part 11: Electronic Signatures and Records in Analytical Laboratories, part 1, *Biopharm*, **12**(11), 28–34, 1999.

17. W. Winter and L. Huber, Implementing 21CFR Part 11: Electronic Signatures and Records in Analytical Laboratories, part 2, Security Aspects for Systems and Applications, *BioPharm* **13**(1), 44–50, 2000.

18. W. Winter and L. Huber, Implementing 21CFR Part 11: Electronic Signatures and Records in Analytical Laboratories, part 3, Data Security and Data Integrity, *BioPharm*, **13**(3), 45–49, 2000.
19. L. Huber and W. Winter, Implementing 21CFR Part 11: electronic Signatures and Records in Analytical Laboratories, part 4, Long Term Archiving and Ready Retrieval, *BioPharm*, **13**(9), 52–57, 2000.
20. W. Winter and L. Huber, Implementing 21CFR Part 11: Electronic Signatures and Records in Analytical Laboratories, part 5, Importance of Instrument Control, *BioPharm*, **13**(9), 2000.
21. W. Winter and L. Huber, Implementing 21CFR Part 11: Electronic Signatures and Records in Analytical Laboratories: Biometrics-Based Authentication—Limitations and Possibilities, Advanstar Communications, *Biopharm*, Suppl., pp. 40–43, Nov. 2000.
22. FDA inspection observations and warning letters, *www.fdawarningletter.com*.
23. PMA's Computer System Validation Committee, Validation concepts for computer systems used in the manufacture of drug products, *Pharm. Technol.*, **10**(5), 24–34, May 1986.
24. FDA guidance for industry Part 11: Electronic Records and Electronic Signature Scope and Applications, draft February 2003, final version August 2003.

18

VALIDATION OF EXCEL SPREADSHEET

HEIKO BRUNNER, PH.D.
Eli Lilly Germany

18.1 INTRODUCTION

Excel is one of the most commonly used computer programs to perform automatic or semiautomatic calculation and visualization of data. Other applications include, but are not limited to, testing of acceptance criteria and using spreadsheets as a view tool for databases. The use of Excel spreadsheets in a regulated environment (cGXP = GMP, GLP, GCP) has to be controlled and validated. Also, periodic revalidation after defined intervals or changes has to be performed. This is consistent with EC guides on Good Manufacturing Practice (chapter 6.7, "Documentation" [1] and chapter 6.16, "Testing" [2]) and the FDA 21 CFR Part 11 [3] electronic records and electronic signatures regulation. A systematic approach in the generation and validation of the spreadsheets application will be required to ensure the validity of the data.

Spreadsheets that can be used for multiple purposes or that can generate multiple types of information are sometimes preferred over single-purpose spreadsheets. An example is a spreadsheet designed to work for different products or a spreadsheet designed to produce multiple result sets from a single data set [e.g., a content uniformity evaluation according to *European Pharmacopoeia* (Pharm. Eur.), *United States Pharmacopoeia* (USP), and *Japanese Pharmacopoeia* (JP) (prior to ICH harmonization)].

Analytical Method Validation and Instrument Performance Verification, Edited by Chung Chow Chan, Herman Lam, Y. C. Lee, and Xue-Ming Zhang
ISBN 0-471-25953-5 Copyright © 2004 John Wiley & Sons, Inc.

There is a negative side to highly flexible spreadsheets. Development and validation of such spreadsheets require substantial time. A spreadsheet designed for a single task may not be as powerful as a more flexible spreadsheet, but it is much easier to create and validate. In this chapter we focus on the validation of spreadsheets generated with Excel but do not cover spreadsheets generated with other software programs. Nevertheless, general strategies and procedures will be applicable for the validation of spreadsheets generated with other programs.

Whenever a standard calculation has to be repeated over a long period of time, the generation of a validated spreadsheet should be considered. Validation of spreadsheets can be time- and resource-intensive. Therefore, it has to be decided if a validated Excel spreadsheet is the solution of choice. Beside the actual validation of each spreadsheet, a procedure has to be established to handle and work with the spreadsheet. GXP-related questions to address include the use of secure data storage locations, missing ER/ES compliance, and audit trail in Excel. These areas are as important as the validation itself and will require resources.

18.2 SCOPE OF CHAPTER

Since Excel is a powerful tool used widely for many different purposes with many options; not all options can be discussed in this chapter. The focus of the chapter is on manually self-created spreadsheets for data calculation and checks against acceptance criteria (logical operations). Excel spreadsheets that are used with other electronic systems for automatic data or information entry, for further operations, or used as a "view tool" for databases are not within the scope of this chapter. Nevertheless, these types of spreadsheets are viewed as a "normal" spreadsheet with automatic entry, and validation, including validation of the interface, will be included to cover this item. In this chapter we provide guidance in validation and revalidation of Excel spreadsheets and information about managing validated spreadsheets.

During the development (design) of a spreadsheet, a clear purpose and data structure are very important for the validation and subsequent application. The design of the spreadsheet should be well ordered, whether data are populated in the spreadsheet automatically by a system or manually by an operator. The spreadsheet should provide a good overview concerning its intent (what is the intent of the spreadsheet—what does it do?), operations (e.g., limit checks, transfer of data/information to other spreadsheets, online generation of information/graphs, use of macros), and calculations. For example, if a spreadsheet uses compound or product-specific values (e.g., acceptance criteria, specifications), these values should be put into a specific section. Formulas using these values for calculation of results should use individual cells. This approach ensures that (1) the spreadsheet is flexible to use for other products, (2) the spreadsheet is flexible in use if the values will change, (3) the spreadsheet is well ordered and values used for result calculations are easily visible, and (4) validation and revalidation can be performed with ease. Cells containing important and critical values have to be protected against change after data entry.

The validation of spreadsheets should ensure that all data generated and all operations and checks performed with the spreadsheet, as well as online-generated graphs, are valid. Depending on individual company policy, the validation approach does not necessarily need to demonstrate that standard calculations (basic arithmetical operations) performed by Excel functions are performing correctly. In addition, Excel software claims and assumes to be validated. This also includes the transfer of data from a table into a chart. However, during validation it has to be demonstrated that the correct connection between data column and graph is established. Figure 18.1 shows an example of a spreadsheet designed for

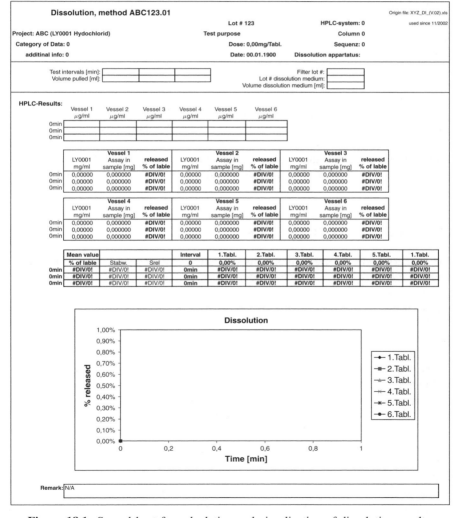

Figure 18.1. Spreadsheet for calculation and visualization of dissolution results.

dissolution testing. The spreadsheet can be used for several products. Results of each individual tablet can be tracked. Additional information concerning the test is given in the header. The online-generated graph helps to visualize similarity of data and curves.

Calculations using several standard calculation functions within one formula or using macros also have to be validated. Macros generated with Excel are a small, self-written software using Visual Basic. Validation of macros should follow the general validation approach for software. Reduction in the extent of validation might be possible.

Excel tests using logical and Boolean operations (e.g., if–then combinations for checking of data against specifications, acceptance criteria, etc.) have to be checked for correct functionality during validation. Excel has tools that are helpful during validation of spreadsheets [e.g., printout of the spreadsheet using different view options (e.g., formula view or view of protected cells)] or the detective tool to track connected or linked cells (Section 18.3.4). Spreadsheets that are used for one time only and are not stored for later use do not need to be validated but checked carefully if used for cGXP operations.

In general, two areas have to be differentiated in establishing (and handling) validated spreadsheets: (1) the development and validation of the spreadsheet itself, and (2) the handling of the spreadsheets, which requires overcoming ER/ES and audit trail issues and securing places/folders for storage of the templates and filled-out spreadsheets. In a network environment this means that the network administrator has to track individual rights and access options for different folders over time.

18.3   VALIDATION PRACTICES

The primary focus during validation is validation of the content of the spreadsheet itself. Nevertheless, items such as location of storage, security of cells and file, as well as access to the file have to be taken into account. The general validation approach for spreadsheets is not different from other validation approaches and can be arranged in three general sections: (1) validation plan, (2) execution of the plan, and (3) validation documentation (protocol).

Validation of spreadsheets should be differentiated in separate sections or documents. These sections or documents should be described in the validation plan and will include the following steps. Some of the steps are used for documentation of the status of the spreadsheet, while others deal with the testing of the spreadsheet. The following 10 items will be differentiated in the validation of spreadsheets.

In this chapter the validation approach will be explained using a simple and complex example. The simple example is a spreadsheet developed and validated for content uniformity determination by UV assay, while the second one is a routine HPLC potency and related substances determination. The simple example is placed directly into the validation section to further illustrate the validation approach, while the second example is located at the end of the chapter for further reading. Different frames will mark descriptions of these examples.

18.3.1 Validation Plan

Prior to validation of a spreadsheet or a combination of spreadsheets (single- or multiple-application spreadsheet), a validation plan document should be put together. The validation plan will include information about its purpose, project background, scope, user characteristics groups involved, and roles and responsibilities. The general information includes:

- File name of spreadsheet, including version number of the spreadsheet
- Storage location of the spreadsheet
- Individual spreadsheets in the file and their interactions or connections (if applicable)
- Operating system and software [e.g., Windows 2000 version 5.0, Microsoft Excel 2000 (9.0.4402 SR-1), including service release pack version]
- Purpose of the spreadsheet (e.g., calculation of linear regression, including equation, graph, and formula used)
- If validation is done retrospectively, data that had been generated with the spreadsheet prior to validation and the date since the spreadsheet was used
- Names of user(s) who developed the spreadsheet and the validation tests
- Extent of validation (Sections 18.3.2, Requirements, to 18.3.10, Review and Approval)

According to cGXP requirements, this plan should be reviewed and approved. Quality assurance involvement is recommended. It would be possible to generate a combined plan/protocol document that includes all information described in this chapter. The actual results generated during validation have to be added to complete the protocol.

An example of a validation plan, including related information, is given in Figure 18.2. This example is used in all individual sections following it to explain what is needed in the various sections. The example is based on a combination of spreadsheets for calculation of content uniformity by UV analysis developed for a development area. The spreadsheet is developed so that it is flexible and can be used for different products. Furthermore, it checks acceptance criteria set for calibration standards and a system suitability test.

18.3.2 Requirements

The requirement section has to describe the requirements of the spreadsheet clearly. The requirements have to be chosen in a manner that permits tests to be developed and performed to check if the spreadsheet fulfills its functions. For each requirement a corresponding test should be performed. The requirement section will include information about purpose, functional, performance, security, and design requirements.

This section of the validation plan should describe the expected function and security of the spreadsheet. The requirements should be formulated in a manner

Company Name
Validation of Excel Spreadsheet
Template Content Uniformity by UV (Version.01)

The first few pages of the document will include a cover page and table of contents with attachments.

Validation Plan
This validation plan describes the prospective validation of the file: Template Content Uniformity by UV (V.01).xls. The file is located at Drive:E\Validated Excel templates\Template_CU_UV_(V.01).xls.

Note: The spreadsheet should be stored in a secure place after development and prior to validation to avoid unauthorized changes during and after validation. The folder should contain only spreadsheets that are currently under validation. When the validation is performed successfully, the file should be moved to a different secure folder. This movement has to be documented.

The spreadsheet contains the following sheets:

- Instructions/Training
- Cover sheet (includes general information concerning sample, operator, equipment, analytical method)
- System suitability for UV equipment
- Content uniformity calculation

The operating system used is Windows 2000 Professional, version 5.0, the application software is MS Excel 2000 SR-1. Validation was performed using a Compaq Deskpro EN, serial number..., containing a... processor.
The individual sheets are related to each other. They will be used for product/project independent calculation of content uniformity according to USP, EP, and JP (slightly different calculation). The individual sheets are:

(a) *Instructions/training*: operation instructions for use of the spreadsheet
(b) *Cover sheet*: The sheet is used to fill in general information concerning project, dose/concentration, sample, equipment, and operator. This information will be transferred automatically to the header of the other sheets to ensure that general sample information is available on each individual sheet.
(c) *System suitability*: Calculation sheet for acceptance criteria concerning calibration standard and control samples.
(d) *Content uniformity calculation*: Calculation sheet for content uniformity according to USP, EP, and JP. Checking of results (potency and relative standard deviation) against acceptance criteria of the individual pharmacopoeias.

The spreadsheet will be used in department/group/area ABC. The spreadsheet will be used after successful validation (or: the spreadsheet is in use since [DATE], if validation is performed retrospectively). John Doe developed the spreadsheet. Jim Doe will perform the validation. Validation will include items... [define which sections (e.g., requirements, design, tests) will be applicable to the validation].

Figure 18.2

that permits them to be checked during validation. For a spreadsheet calculating the linear regression, the following parameters are described:

- Minimum number of entries to be included (e.g., 10)
- Automatic calculation of slope and y-intercept of the linear regression according to the equation used
- Automatic calculation of the standard deviation according to the equation used
- Automatic check to determine if the y-intercept and the standard deviation meet defined acceptance criteria
- Protection of all cells where data should not be entered

A detailed example for a spreadsheet used for content uniformity calculation is given in Figure 18.3.

Requirements
The example contains a combination of spreadsheets (a to d; see the validation plan example). Each spreadsheet contains colored cells that indicate where data entry will be performed. All other cells are colorless and protected against information entry and update. Data format and cell protection will be documented in attachments. The requirements for the individual spreadsheets are described below:

(a) Instructions/Training
This sheet contains general information concerning the development and use of the spreadsheet. It describes the connection of cells of the various sheets and contains information concerning actions to be taken in redevelopment/update of the spreadsheet (e.g., update of limits and acceptance criteria, data format or formulas).
   No special training recommendations or user tips are necessary for this spreadsheet.

(b) Cover sheet
The following information will be put into the spreadsheet:
   General information: Product, including description (e.g., tablet, dose); method name, including revision and technique (e.g., UV); project; information concerning sample(s); date of test and lot number; equipment identifier; raw data file name of computerized system; and analyst.
   Information concerning reference/calibration standard: compound, potency, and reference for documentation of standard preparation.
   Information concerning sample(s): number of samples, storage condition, reference to sample preparation documentation.

(c) System suitability
General information will be transferred automatically from the cover sheet. Additional information that will be put into the spreadsheet includes:

- Individual sample code
- Preparation concerning calibration and reference standard

Figure 18.3. (*continued overleaf*)

- Weight of standards and dilution(s)
- Blank sample preparation
- Used solutions or reference to solution preparation
- Equipment used for weighing and dilution

Besides the related analytical information, values for acceptance criteria will be entered into the spreadsheet. Cells containing this information will be protected against changes after data entry. Acceptance criteria will be established for:

- Relative standard deviation (RSD) for reproducibility of standards
- Maximal difference between calibration and reference standard
- Maximal difference between n and $n + 1$ calibration
- Maximal absorbance for blank sample

The spreadsheet will calculate and assess:

- Calculation of concentration of standard stock solution in mg/mL
- Calculation of diluted standard solutions in μg/mL
- Calculation of absorbance per unit weight for calculation of standard reproducibility
- Calculation of reproducibility for calibration and control standard and assessment against predefined acceptance criteria
- Assessment for blank run
- Calculation of relative standard deviation for standards and assessment against predefined acceptance criteria
- Calculation of mean absorbance of standards
- Calculation of absorbance per microgram of standard
- Calculation of mean value of absorbance per microgram
- Calculation of mean of control standards as percent of label claim and assessment against predefined acceptance criteria
- Calculation of mean for control standards at the end of analytical test series as percent of control standard at the beginning of the test series and assessment against the predefined acceptance criteria

(d) Content uniformity calculation

General information will be transferred automatically from the cover and system suitability sheet. This spreadsheet calculates and evaluates the following parameter:

- Recalculation of USP acceptance criteria from percent of label claim in label claim/label weight
- Recalculation of EP acceptance criteria from percent of mean value in percent of label claim
- Calculation of potency of samples as potency per label weight and percent of label claim
- Calculation of mean potency value, standard deviation, and relative standard deviation for the sample set

Figure 18.3. (*continued*)

- Display lowest and highest potency value
- Assessment of content uniformity according to USP predefined criteria
- Assessment of content uniformity according to EP predefined criteria
- Assessment of content uniformity according to JP predefined criteria

For the calculation this spreadsheet will use information from other spreadsheets:

- Weighing of standard
- Potency of standard
- Volume of standard solution
- Absorbance of standard (mean value)
- Absorbance of blank

Furthermore, the documentation has to describe which formulas are used in the spreadsheet for the calculation and explanation of abbreviations; for example:

- System suitability test:

RSD: $\dfrac{s \times 100}{\overline{x}}$

Mean value: $\dfrac{\sum .x_i}{n}$

All spreadsheets contain information concerning origin of template and the date when they were put into use.

Figure 18.3. (*continued*)

18.3.3 Design

The design section will include information about purpose, system and functional overview, data flow, data sheet formats, and test strategy. The section describes how the requirements are addressed in the spreadsheet. The main part of the design documentation contains a printout of the spreadsheet in the "formula" view, with line and column headers. The purpose of the individual formulas and/or macro must be described, including the sequence in which the calculation will be performed. In addition, the following information should be provided:

- The cells content or formulas that are essential to fulfill the requirements and therefore will be checked during validation
- The cells that will be protected (could be performed by highlighting these cells)

Excel has features available that can assist during validation. One of these features allows a printout of protected individual or grouped cells as well as number formats. Figure 18.4 is an example.

Design

The design describes the transformation of the requirements in the spreadsheet. In most cases a printout of the spreadsheet in the formulas view will be appropriate for addressing the transformation of requirements.

Types of cells used for the requirements are as follows:

(a) All cells that are used for data and/or information input
(b) All cells that are used for calculations
(c) Cells that are cross-linked to other spreadsheets

Testing will include all types of cells mentioned in (a) to (c). All cells that will not be used for data entry are protected. A printout of the spreadsheets documenting the protected cells will be provided.

Figure 18.4

18.3.4 Tests

The test chapter section will include information about purpose, test documentation, review and approval, formal test plan, and test execution. Testing the spreadsheet should take all requirements into account. Tests using known cases and data including known results have to be developed. The data set used for testing should include representative data. If applicable, tests should also include data sets that will test values that exceed the anticipated range of normal operations and check for data close to the acceptance criteria limits. Data that make no sense (text in a numeric field) or are not allowed to be used in the calculation should be considered in the testing phase as well. The simplest test to validate a spreadsheet calculation is to perform a manual calculation and document it. All tests performed, results expected, acceptance criteria for the tests, and procedures for alternative calculation (e.g., manual calculation) will have to be documented. Persons who perform the testing have to be trained appropriately and must be mentioned in the documentation.

Excel offers tools to support validation and documentation approaches (e.g., the cross-references of cells could be visualized by the "detective" tool under "Extras", "Detective"). Several options are available. Figure 18.5 illustrates the importance of this tool. The data set that is used for the validation gave the same

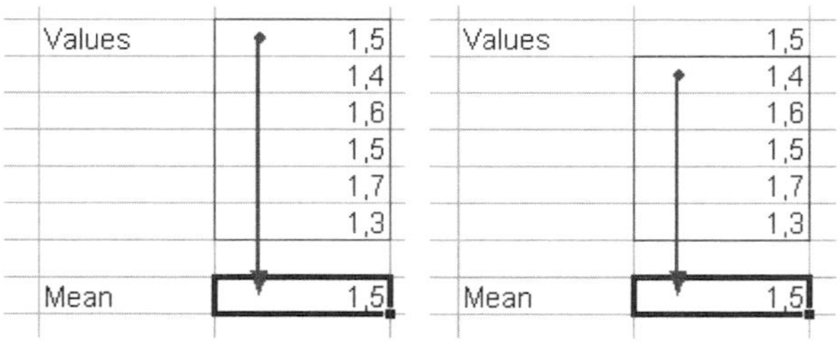

Figure 18.5. Using the "detective" tool of Excel to assist validation.

Tests

For a check of the spreadsheets, test data were used. The test data represent values close to the acceptance criteria, which are within and outside the acceptance range. The following checks will be performed:

(a) The assessment of the acceptance criteria function
(b) The correct calculation of results

Test 1: correct calculation of content uniformity results

Test 2: correct data format and rounding to predefined significant places

Test 3: test values close to but within acceptance criteria

Test 4: test values close to but outside acceptance criteria

Test 5: test for a combination of acceptance criteria (e.g., system suitability test)

Test 6: check for the correct cell connection used for information transfer

Test 7: security test

Note: Tests have to be performed for each acceptance criterion. If several acceptance criteria have to be met to get overall acceptance (e.g., system suitability test: relative standard deviation of standard absorbance and blank sample with no absorption), the overall functionality has to be checked.

Figure 18.6

mean value. However, in the case on the right, the calculation is missing the first value (1.5). The error that excludes the first cell in the calculation can be visualized with the detective tool as demonstrated.

Testing must be documented according to cGXP requirements. The results have to be reviewed and approved. Documentation of the entire testing procedure should have a printout of all results obtained with the spreadsheet as well as results obtained with the alternative calculation. Figure 18.6 is an example.

18.3.5 Security

The security description will include information about physical and logical security. Security procedures have to be documented. Excel and the Windows environment offers several options. The following possibilities should be taken into account:

- Use of passwords for cells, spreadsheets, and/or a combination of spread-sheets (*Note:* A printout of the spreadsheet showing the protection of cells should be made)
- Procedure for allocation of passwords

Security

The spreadsheet is stored as a password-protected file in the folder: E\Validated Excel templates\. The file name is "Template_CU_UV_(V.01).xls."

The following people (group or department) have access (read only) to the folder:

- Lisa
- Tim
- Tom

Note: Detailed documentation according to the access levels for the folder should be provided and kept up to date (e.g., John Doe knows the password).

Figure 18.7

- Saving the (master) spreadsheet only on secure drives where only authorized people have access
- Saving the spreadsheet as a sample model (.xlt) (this could only be performed if local use of the spreadsheet is intended)
- Making a backup

The security strategy selected has to ensure that cells containing formulas (on purpose, by mistake, or by the auto-save function) of the spreadsheet cannot be overwritten. In today's standard office network environment, in some cases, the network itself is, not validated and does not fulfill the electronic record/electronic signature requirement. Therefore, the validated spreadsheets should be stored in a protected drive to which only restricted personnel have access. Furthermore, the server used for storage/handling should be qualified. Figure 18.7 is an example.

18.3.6 Documentation for User and Training

The training description will include information about purpose, responsibilities, training deliverables, and training records. In some cases it will require generation of a document explaining the use of the spreadsheet. In this document, the function of the spreadsheet, the formulae used, the location of the secure templates, and the file name and path to use to save the completed form should be included. Training has to be documented. Figure 18.8 is an example.

Documentation for User and Training

The spreadsheet is stored as a password-protected file in the folder E\validated Excel pattern\...\."Template_CU_UV_(V.01).xls."

The following people have access (read only) to the folder. (*Note*: A detailed documentation according to access levels for the folder should be provided and kept up to date.) John Doe knows the password for the protection of the spreadsheet.

Figure 18.8

The spreadsheet does not require special training. It is designed as a "fill out only" spreadsheet. The completed spreadsheet will be stored in folder E\...\. After review, second-person verification, and sign-off the electronic version will be archived in database ABC.

Figure 18.8. (*continued*)

18.3.7 Changes and Version Control

The changes and version control section will include information concerning purpose, responsibilities, and testing. Changes in the operating system and/or in Excel (e.g., new version and other significant changes in the spreadsheet) have to be documented, reviewed, and approved. The documentation will include:

- Description of the changes
- Person who carried out the change
- Date when change was finalized
- Person who performed testing after the change
- Date of last test

The spreadsheets will have a unique identifier (e.g., file name and version number). All changes of the spreadsheet have to be documented in a document history. An example is shown in Table 18.1.

Every significant change including use of different language versions of Excel will require revalidation of the spreadsheet. Table 18.2 provides examples when revalidation (testing) of the spreadsheet is required. Establishing changes in the spreadsheet will follow the change control process. Figure 18.9 is an example.

Changes and Version Control
Every change in operating system, the application software (Excel), and significant changes in the spreadsheet will be documented, reviewed, and approved. The documentation will contain:

- Description of the change
- Name of the person performing the change
- Date of the change
- Name of the person performing the testing
- Date when the tests were performed

A change control history is provided.

Figure 18.9

Table 18.1. Example of Spreadsheet History

Description	Name/Date
Windows 95, Version 4.0	Initial configuration
Microsoft Excel®, Version 7.0	Initial configuration
ABC calculation Version 1.xls	Initial configuration
Windows NT, Version 2.0	Doe, 04. March 1999
ABC calculation Version 2.xls	Doe, 02. June 2000

Table 18.2. Examples for Changes and Testing Requirement

Change	Testing
Change to a formula or macro	Testing required
Addition or deletion of a formula or macro	Testing required
Excel software upgrade	Testing required
Operating system upgrade	Testing required
Relocation of desktop computer	Testing not required
Change to computer CPU	Testing not required
Change to computer monitor	Testing not required
Change to computer keyboard	Testing not required
Change to computer mouse	Testing not required

18.3.8   Archiving

The archiving section will include information about purpose, record classification, and archive location(s). It has to be defined which data generated during the validation of the spreadsheet will be classified as raw data or results (GXP records): the electronic data or a printout. The raw data and all data and documents that have been generated during validation must be archived. Figure 18.10 is an example.

Archiving
The following documents will be archived:

- Validation plan and protocol
- Raw data

A definition of raw data is provided in the procedure.... Responsibility and how to proceed for archiving is described in the procedure....

Figure 18.10

Periodic Review
The password-protected file is stored in a secure folder. Nevertheless, a periodic review concerning safety and functionality will be made at least every 12 months.

Figure 18.11

18.3.9   Periodic Review (Revalidation)

The periodic review statement will include information about purpose, process, and intervals. This section will define the interval of periodic review. The test(s) of functions of the spreadsheet should be checked during periodic review. The planned tests, expected results, and acceptance criteria have to be specified. Specific data sets for the tests should be generated. Figure 18.11 is an example.

18.3.10   Review and Approval

The review and approval (system acceptance) will include information about the approvers. All validation documents have to be reviewed and approved with involvement of the quality assurance unit. Figure 18.12 is an example.

In addition to the foregoing items that have to be covered, there are other topics that might need to be taken into account. These topics include (note that several of these topics are related to computer system validation):

- Generation of a master document list
- Generate inventory (list) for validated spreadsheets that are in use
- System retirement
- Business contingency plan when spreadsheet is unavailable (e.g., due to network issues)

Review and Approval
The multiapplication spreadsheets in the file ABC (V.01) were tested successfully and comply with the requirements. The spreadsheets can be used without limitations. The individual spreadsheets are:

- Instructions
- Cover sheet
- System suitability
- Content uniformity calculation

Figure 18.12

- Validation master plan (e.g., if several similar spreadsheets should be validated)
- Vendor assessment (needed only if a vendor delivers a validated spreadsheet)
- Procedure(s) and documentation of training for users

A more detailed example of a spreadsheet validated according to the description given in this chapter is presented in Figure 18.13. This example is a multiapplication spreadsheet for product-independent calculation of potency and related substances.

Company Name
Validation of Excel Spreadsheet
Template Potency and Related Substances (Version.01)

Validation Plan
This validation plan describes the prospective validation of the file: Template Potency related substances (V.01).xls. The file is located at Drive:E\Validated Excel templates\....
 The spreadsheet contains the following sheets:

- Instructions/training cover sheet (includes general information concerning sample, operator, equipment, analytical method)
- System suitability (for potency determination)
- System suitability calibration standard 2 (for related substances determination)
- Potency calculation
- Related substances (rel sub) calculation

The operating system is Windows 2000 Professional, version 5.0. The application software is MS Excel 2000 SR-1. Validation was performed using a Compaq Deskpro EN, serial number, processor.
 The individual sheets are connected to the result summary sheet and will be used for product-independent calculation. The individual sheets are:

(a) *Instructions/training:* Operation instructions for use of the spreadsheet
(b) *Cover sheet:* Used to enter general information concerning project, dose/concentration, sample, equipment, and operator that will automatically be transferred to other sheets.
(c) *System suitability (SysSuit):* Calculation sheet for acceptance criteria concerning calibration standard, control samples, and other items (e.g., tailing, resolution)
(d) *SysSuit calibration standard 2:* (e.g., calibration standard for rel subs): Calculation of acceptance criteria for second calibration standard
(e) *Potency:* Calculation sheet for potency determination (e.g., two or three preparations per sample, one or two injections per sample)

Figure 18.13

(f) *Related substances (rel subs):* Calculation sheet for related substances determination (e.g., two or three preparations per sample, and one or two injections per sample)

The spreadsheet will be used in department/group/area ABC.

The spreadsheet will be used after successful validation (or: the spreadsheet is in use since [DATE], if validation is performed retrospectively).

The spreadsheet was developed by John Doe. The validation will be performed by Jim Doe.

Validation will include items 18.3.2, Requirements, to 18.3.9, Periodic Review *(see relevant sections of this chapter).*

Requirements

The example contains a combination of spreadsheets [sections (a) to (f) in the validation plan]. The requirements for the individual spreadsheets are as follows:

(a) Instructions/Training

This sheet contains general information concerning the development and use of the spreadsheet. It describes the connection of cells of the various sheets and contains information concerning actions to be taken in redevelopment/update of the spreadsheet (e.g., update of limits and acceptance criteria, data format, or formulas).

No special training recommendation or user tips are necessary for this section of the spreadsheet.

(b) Cover sheet

The following information will be entered into the spreadsheet:

General information: Product, including description (e.g., tablet, dose); method name, including revision and technique (e.g., HPLC); project information concerning sample(s) (e.g., stability samples three months); date of test and lot number; equipment identifier; column; raw data file name of computerized system and analyst. This general information will be transferred to each spreadsheet.

Information concerning reference/calibration standard: Compound(s), potency, reference for documentation of standard preparation.

Information concerning sample(s): number of samples, storage condition, reference to sample preparation documentation.

(c) System suitability

The following information will be entered into the spreadsheet:

- Solution(s) used (e.g., reference lot, supplier, preparation, volumes, dilutions, weighings, references to solution preparation if applicable)
- Injection volume
- Equipment used for weighing and dilution
- Solution for blank run
- Solution for sensitivity test
- Decisions if acceptance criteria for control standard should be used (e.g., in percent of label claim, percent change at end of sequence)

Figure 18.13. *(continued overleaf)*

- Acceptance criteria for tailing factor, percent injection reproducibility, relative standard deviation of calibration and/or control standard

The spreadsheet will calculate and assess:

- Calculation of concentration of standard stock solution in miligram/mL
- Calculation of diluted standard solutions in microgram/mL
- Assessment of USP tailing factor using predefined acceptance criterion
- Calculation of peak area versus weighing for injection reproducibility
- Calculation of injection reproducibility for calibration and control standard and assessment against predefined acceptance criteria
- Assessment of blank run
- Calculation of relative standard deviation for standard runs (peak areas) and assessment against predefined acceptance criteria
- Calculation of mean for peak areas of standards
- Calculation of peak area per microgram of standard
- Calculation of mean value of peak area per microgram
- Calculation of mean of control standards as percent of label claim and assessment against predefined acceptance criteria
- Calculation of mean for control standards at the end of sequence as percent of control standard at beginning of sequence and assessment against predefined acceptance criteria

(d) System suitability of second standard
If a second standard (e.g., diluted standard of "100% standard" for calculation of related substances) is used in the method, a separate spreadsheet should be used for calculation and assessment of acceptance criteria concerning this standard.

(e) Calculation of the potency samples
For the flexibility of the entire spreadsheet, the spreadsheet contains calculations for various options:

- Three samples, one injection per sample
- Three samples, two injections per sample
- Two samples, one injection per sample
- Two samples, two injections per sample

The following information will be entered into the spreadsheet:

- Batch number
- Storage condition (if samples are from a stability study)
- Weighing of sample with choice of unit (e.g., mg or g)
- Mass of the composite of the grinded dosage units with a choice of unit
- Unit for results from the chromatogram
- Choice if calculation will be performed for defined dose strengths

Figure 18.13. (*continued*)

- Values for acceptance criteria, RSD, upper and lower specification limits, including significant figures
- Significant figures of results

The spreadsheet calculates and evaluates the following parameters:

- Recalculation of result/weighing in results/mean mass
- Calculation of mean result/product for each sample
- Calculation of result in percent of label claim for each sample
- Calculation of mean of mass/product and in percent of label claim
- Calculation of RSD of potency results
- Conversion of acceptance criterion in percent of label claim in mass/product
- Assessment of RSD with acceptance criteria
- Assessment of percent label claim with acceptance criteria

(f) Calculation of related substances samples
For the flexibility of the entire spreadsheet, the spreadsheet contains calculations for various options:

- Three samples, one injection per sample
- Three samples, two injections per sample
- Two samples, one injection per sample
- Two samples, two injections per sample

The following information will be put into the spreadsheet:

- Hold time
- Retention time of active pharmaceutical ingredient (API) peak
- Unit for related substances calculation
- Choice if calculation will be performed for defined dose strengths
- Mass unit of API potency
- Decision if calculation should be performed for label claim potency
- Selection for further calculations: percent of label claim, percent of actual API content, percent of sum of peaks
- Peak name
- Retention time of related substance peak
- Input of required significant figures for results of single related substance, mean value of related substances, related substances calculated with respect to label claim
- Input of acceptance criteria (value, significant figures): related substances

The spreadsheet calculates and assesses the following parameters:

- Relative retention time
- Calculation of results from mass/weighing to mass/average mass
- Recalculation from mass/average mass to mass/label claim potency
- Calculation of related substances result as percent of label claim
- Calculation of related substances results in percent of API potency

Figure 18.13. (*continued overleaf*)

- Calculation of related substances results in percent of API potency and sum of related substances
- Calculation of total (sum of) related substances
- Calculation of largest individual related substance
- Individual results against acceptance criteria, if defined (e.g., ICH limits)

Furthermore, which formulas are used for the calculations and an explanation of abbreviations used must be documented; for example:

- System suitability test:

$$\text{standard deviations: } \sqrt{\frac{\sum(x_i - \bar{x})^2}{n-1}} \qquad \text{RSD: } \frac{s \times 100}{\bar{x}}$$

$$\text{mean value: } \frac{\sum x_i}{n}$$

- Abbreviations used in formulas for example:

$$a: \qquad \text{potency of reference standard}$$

$$c_x: \qquad \text{concentration of solution } x$$

Cells that will be used for data entry are colorless; protected cells are colored.

All other validation sections, such as Design, Test, Security, Documentation for User and Training, Changes and Version Control, Archiving, Periodic Review, and Review and Approval, can be taken from the content uniformity example described earlier in the chapter.

Figure 18.13. (*continued*)

18.4 COMMON PROBLEMS AND SOLUTIONS

Many different problems and issues can occur in establishing or using validated spreadsheets. Some examples of theses issues are discussed below. In a standard office environment, most personal computers (PCs) are not validated. Also, the computer network (see Section 18.3.5) is not fully under control. Excel has some cGXP compliance issues. The developer and user of spreadsheets should be aware of these issues and develop procedures to overcome and close the gaps. The most critical issue is that Excel is not compliant according to electronic record and electronic signature requirements and offers no audit trail (for details see 1 and 2). Third-party solutions are available to address these issues [4]. Besides validation, the general handling according to cGXP requirements is the most common issue.

1. *Electronic record/electronic signature.* Currently, requirements concerning ER/ES according to FDA 21 CFR Part 11 cannot be implemented with Excel spreadsheets. Concerns include file protection, security, user access, audit trail, version retention, file deletion or erasure, electronic signature(s), and network issues. These can be resolved using a hybrid system, an electronic version of the spreadsheet and an approved hard copy. After data entry (and subsequent to

saving the document permanently) a printout of the spreadsheet is made, which is reviewed, approved, and archived.

The review process of filled-out spreadsheets will be in question from a regulation perspective. Incorrectly completed spreadsheets will normally be revised and reprinted. The first, incorrect version will be destroyed. An audit trail of the review process is not possible. This can be overcome by keeping the incorrect version(s). Another system to track the review process is the introduction of check boxes close to data-entry cells or columns on the spreadsheet that will demonstrate the correctness of the entry. It should be noted that the requirements for ER/ES compliance and for an appropriate audit trail is needed only if a spreadsheet is saved. If the PC is used as a typewriter, the requirements are not applicable.

2. *Audit trail.* Excel does not offer an audit trail according to cGXP requirements. Using the "share workbook" and "track changes" functions could establish a kind of audit trail. The change-history setting can be set to keep the change history for up to 32,767 days. Nevertheless, a system has to be established to overcome this problem. A solution would be to import the spreadsheets electronically into a cGXP-compliant database (document management system). After such importation, the spreadsheet version stored on a nonqualified drive should be deleted.

3. *Change of software version(s).* Changes in software version(s) require revalidation of the spreadsheet irrespective of whether a new version of Excel or Windows is used. It will be a business decision to stay with the old software version to avoid revalidation and save the resources required. In a large company, this will not be an easy decision. In this case, care should be taken to ensure the use of the appropriate software and spreadsheet version, especially if spreadsheets are used in a shared manner. In the case of a global rollout of new software, revalidation must take place prior to the rollout to ensure the use of validated spreadsheets starting on the day of the rollout.

There are generally two options that can be followed for revalidation. A spreadsheet developed and validated for Excel 97 could be opened with Excel 2000, revalidated and saved as an Excel 97 file. It can also be opened with Excel 2000, saved as an Excel 2000 file, and followed by revalidation. The decision as to how the spreadsheet is revalidated and its subsequent use must be documented.

4. *International use of validated spreadsheets.* As described above, the validation of spreadsheets needs to include the software version(s) of Excel and Windows. A country-specific setting on the PC or the software has to be addressed, as this setting can cause significant problems. In different language the terms for arithmetical operations are not necessarily identical. Therefore, Excel will not be able to execute the calculation correctly (e.g., the English term for an operation is *convert* and the German term is *Umwandel* using Excel 97). The software is not able to translate this term correctly and therefore the resulting calculation is not correct. Fortunately, Excel 2000 translates this function correctly. This issue also has to be addressed when a new Excel version is used. An additional difference between English and German is the decimal "dot" in English (2.5)

compared with the decimal "comma" in German (2,5). Excel 2000 converts the data correctly. Nevertheless, older versions might present problems, and attention should be given to ensure that a value of 2.5 is converted to 2,5, not to "May 2."

Translations of standard formulas using different Excel language versions should be performed automatically by Excel. Nevertheless, revalidation has to carried out when switching the Excel language. At least the language-specific explanations and training instructions using the text fields have to be translated.

5. *Significant places in calculations and results.* During development and validation of spreadsheets, the appropriate use of significant places should be addressed. Rounding rules for analytical method precision, precision of equipment used, and specifications and acceptance criteria have to be taken into account. If intermediate and final results of the spreadsheet are reported, it should define the number of significant places that Excel used for the individual calculations. Excel might use more significant places as displayed and perform rounding.

6. *Manual versus automatic calculation.* Excel offers two options to perform calculations: manual and automatic (menu "Extras," "Options," "Calculations"). If automatic calculation is selected, calculation will occur immediately after data entry. If manual calculation is selected, calculation/recalculation will occur only after pressing the "F9" button. Manual versus automatic calculation might cause confusion in using validated spreadsheets. If manual calculation is chosen in the original validated spreadsheet, the calculation will not be performed automatically even if automatic calculation is chosen during use. It will have no impact on the correctness calculation. If users are used to automatic calculation, they will report the wrong results, especially if they recalculate by putting in different numbers and do not press "F9."

Spreadsheets can be a powerful tool to perform calculations and data evaluation more efficiently and to reduce errors in a cGXP environment. However, the possible problems and resources needed to develop, validate, and revalidate spreadsheets should not be underestimated.

REFERENCES

1. *Rules and Guidance for Pharmaceutical Manufacturers and Distributors*, part 4, *Guide to Good Manufacturing Practice for Medical Products*, Chapter 6.7, Documentation, 1997.
2. *Rules and Guidance for Pharmaceutical Manufacturers and Distributors*, part 4, *Guide to Good Manufacturing Practice for Medical Products*, Chapter 6.16, Testing, 1997.
3. FDA CFR Part 11, *www.fda.gov/ora/compliance_ref/part11/*.
4. T. T. Phan, Technical consideration for the validation of electronic spreadsheets for complying with 21 CFR Part 11, *Pharm. Technol.*, pp. 50–62, Jan. 2003.

INDEX

Analytical Method Validation and Instrument Performance Verification, Edited by Chung Chow
Chan, Herman Lam, Y. C. Lee, and Xue-Ming Zhang
ISBN 0-471-25953-5   Copyright © 2004 John Wiley & Sons, Inc.